OHS Electronic Management Systems for Construction

Occupational accidents have a massive personal and social cost as well as a major financial one. The construction industry is one of the most dangerous industries, accounting for around 20–30 per cent of all occupational deaths worldwide. The accompanying financial cost is either absorbed directly or passed on in the form of higher insurance costs. In addition, regulatory bodies have started to impose legal accountability on all the parties along the construction supply chain.

OHS is hard to implement. Construction projects are complex, with a fluid workforce, and the regulatory framework is highly elaborate. This book presents a theoretical framework which is designed to overcome these difficulties, integrating OHS management in construction, knowledge management, and web technologies. This framework is explained in a clear step-by-step way, as are features such as a systematically developed corporate safety memory, a virtual learning portal to facilitate on-demand safety training, a groupware for collaborative OHS monitoring, and an online platform for social learning of safety.

The ultimate aim of this book is to aid the development of an established safety culture at the organisational level and the formation of an industry-wide community of safety practice, thereby curtailing construction accidents. This is essential reading for OHS professionals and construction managers attempting to change their industry for the better, as well as advanced students and researchers.

Imriyas Kamardeen is a Senior Lecturer at the University of New South Wales, Australia, and a former cost consultant.

Spon Research

Spon Research publishes a stream of advanced books for built environment researchers and professionals from one of the world's leading publishers. The ISSN for the Spon Research programme is ISSN 1940-7653 and the ISSN for the Spon Research E-book programme is ISSN 1940-8005.

Published:

Free-Standing Tension Structures: From tensegrity systems to cable-strut systems
978-0-415-33595-9
W. B. Bing

Performance-Based Optimization of Structures: Theory and applications
978-0-415-33594-2
Q. Q. Liang

Microstructure of Smectite Clays and Engineering Performance
978-0-415-36863-6
R. Pusch and R. Yong

Procurement in the Construction Industry: The impact and cost of alternative market and supply processes
978-0-415-39560-1
W. Hughes et al.

Communication in Construction Teams
978-0-415-36619-9
S. Emmitt and C. Gorse

Concurrent Engineering in Construction Projects
978-0-415-39488-8
C. Anumba, J. Kamara and A.-F. Cutting-Decelle

People and Culture in Construction
978-0-415-34870-6
A. Dainty, S. Green and B. Bagilhole

Very Large Floating Structures
978-0-415-41953-6
C. M. Wang, E. Watanabe and T. Utsunomiya

Tropical Urban Heat Islands: Climate, Buildings and Greenery
978-0-415-41104-2
N. H. Wong and C. Yu

Innovation in Small Construction Firms
978-0-415-39390-4
P. Barrett, M. Sexton and A. Lee

Construction Supply Chain Economics
978-0-415-40971-1
K. London

Employee Resourcing in the Construction Industry
978-0-415-37163-6
A. Raiden, A. Dainty and R. Neale

Managing Knowledge in the Construction Industry
978-0-415-46344-7
A. Styhre

Collaborative Information Management in Construction
978-0-415-48422-0
G. Shen, A. Baldwin and P. Brandon

Containment of High Level Radioactive and Hazardous Solid Wastes with Clay Barriers
978-0-415-45820-7
R. N. Yong, R. Pusch and M. Nakano

Performance Improvement in Construction Management
978-0-415-54598-3
B. Atkin and J. Borgbrant

Organisational Culture in the Construction Industry
978-0-415-42594-0
V. Coffey

Relational Contracting for Construction Excellence: Principles, Practices and Partnering
978-0-415-46669-1
A. Chan, D. Chan and J. Yeung

Soil Consolidation Analysis
978-0-415-67502-4
J. H. Yin and G. Zhu

OHS Electronic Management Systems for Construction
978-0-415-55371-1
I. Kamardeen

Forthcoming:

FRP-Strengthened Metallic Structures
978-0-415-46821-3
X. L. Zhao

OHS Electronic Management Systems for Construction

Imriyas Kamardeen

Routledge
Taylor & Francis Group

LONDON AND NEW YORK

First published 2013 by Routledge

2 Park Square, Milton Park, Abingdon, Oxfordshire OX14 4RN
52 Vanderbilt Avenue, New York, NY 10017

Routledge is an imprint of the Taylor & Francis Group, an informa business

First issued in paperback 2019

British Library Cataloguing in Publication Data
A catalogue record for this book is available from the British Library

Library of Congress Cataloging-in-Publication Data
Kamardeen, Imriyas.
OHS electronic management systems for construction / Imriyas Kamardeen.
    p. cm.
  Includes bibliographical references and index.
  1. Construction industry – Safety measures. 2. Construction industry –
  Safety regulations. 3. Construction industry – Accidents – Prevention.
  4. Construction industry – Health aspects. I. Title. II. Title: Electronic
  management systems for construction.
  HD7269.B89K36 2013
  363.11´9624–dc23   2012025136

ISBN13: 978-0-415-55371-1 (hbk)
ISBN13: 978-0-367-38038-0 (pbk)

Typeset in Sabon
by HWA Text and Data Management, London

To my parents, and Zakia and Mahdiyya

Contents

Figures

Tables

Preface

The construction industry performs poorly in occupational health and safety (OHS) globally. Despite numerous OHS initiatives and campaigns by policy-makers, authorities, builders, construction unions and researchers, statistics confirm that construction workers continue to be killed or injured at work each year at a higher rate than that of any other industry sector. The persistent poor safety record is frequently attributed to the very nature of construction such as fragmentation, complexities and dynamism of construction processes, budgetary and schedule constraints, skill shortages, multiculturalism and language barriers, multilayered subcontracting, and lack of OHS innovation and organisational learning. There is a compelling need for innovative methods to address these peculiar challenges towards improving the safety performance of the construction industry globally.

Developments in e-business technologies have revolutionised the business operations of many industry sectors; construction is no exception. The deployment of e-business solutions by the construction industry has brought about considerable improvements to its performance. It has enabled the industry to overcome many of its traditional challenges such as fragmentation, information inconsistency, rework, delays, cost overruns, disputes, etc. There is also a growing trend in the adoption of knowledge management systems in construction as a recognised essential element for innovation and process improvement. Nevertheless, the opportunities offered by these paradigms and technologies have not been adequately utilised to revamp OHS management practices in construction. One key reason for this is the scarcity of literature that provides clear directions as to how these technologies may be integrated into OHS management.

In this vein, this book advocates a novel concept of knowledge-based OHS management using web technologies in construction. It intends to explain how construction organisations can implement a knowledge-based OHS management approach with the support of web technologies, in a clear, concise and easy-to-follow fashion. Chapters in the book cover elements of OHS management with new perspectives, instituted by the main theme. The chapters discuss the research and development of improved solutions for

OHS management via a careful integration of knowledge management, OHS and e-business principles.

The book will be a valuable handbook for builders, OHS professionals and developers of online project management software. For builders and OHS professionals, it will provide an understanding of new ways by which potential occupational incidents can be controlled before they onset. For software developers, it shows the ways to improve existing online project management applications with new capabilities. The book will equally be a valuable resource for researchers, academics and tertiary students in the construction space. It will provide insights into how modern information and knowledge management technologies may be applied to address dynamic challenges in the mainstream construction field.

Acknowledgements

Images used on pp. 62–3 are from WorkSafeBC and Vocam videos and used here with permission. For more information on free WorkSafeBC occupational health and safety products visit www.worksafebc.com. The screenshots presented throughout the book are taken from the prototype applications at www.ohsplanner.consafesolutions.net, www.trainer.consafesolutions.net, www.ohscommunity.consafesolutions.net, and http://1.eohsapp.appspot.com/, developed by the author.

This book would not have been a success without the valuable support rendered by various people. I am indebted to express my gratitude to all of these support pillars. I thank my employer, the University of New South Wales (UNSW), Australia, for providing me with resources necessary to carry out the research that underpins this book. Special thanks to professionals in the construction industry of New South Wales, Australia, for their voluntary support and participation in the research. I appreciate my colleagues in the Construction Management and Property Program at UNSW and friends for their encouragement and scholarly advice from time to time.

Last but not least I extend my appreciations to my wife, Zakia, for her love, support and continued patience throughout the development of this book.

Abbreviations

ASCC	Australian Safety and Compensation Council
CFMEU	The Construction, Forestry, Mining and Energy Union
CMS	Content management system
COHSMS	Construction occupational health and safety management system
CoP	Community of practice
CoSP	Community of safety practice
DIAC	Department of Immigration and Citizenship
GDP	Gross domestic product
ICT	Information and communication technologies
ILO	International labour organisation
KM	Knowledge management
MSDS	Material safety data sheet
NSW	New South Wales
OHS	Occupational health and safety
PM	Project manager
PPE	Personal protective equipment
PtD	Prevention through design
PTW	Permit-to-work
SWMS	Safe work method statement
UK	United Kingdom
VCoSP	Virtual community of safety practice
WCR	Weighted consensus rating
WHS	Work health and safety

1 Introduction

Challenges for accident prevention in construction

Accident prevention and control in construction is a persistent global challenge, with construction having one of the worst industrial safety records of all areas, including high-risk industries such as the chemical, mining, electrical and transportation (Hu et al. 2011). In the US, for example, the construction industry accounts for 19 per cent of all occupational fatalities and remains the highest source of fatal occupational accidents (Bureau of Labor Statistics 2010). In the UK, the construction fatality rate constitutes 21.5 per cent of total occupational fatalities (Health and Safety Executive 2010b), while reportable non-fatal injuries averaged 16 per 1,000 workers, significantly higher than the overall average of 10 per 1,000 workers (Labour Force Survey 2009). In Australia, the construction industry's accident rate of 22 per 1,000 workers is much higher than the national rate for all industries of 14 per 1,000 workers. Its fatality rate of 5.9 per 100,000 workers in 2009–10 is almost three times the rate for all industries, which is 1.9 per 100,000 workers (Safe Work Australia 2011). Safe Work Australia has estimated the economic cost of workplace injury and illness to the Australian economy for the 2005–6 financial year to be $57.5 billion, 5.9 per cent of GDP. The construction industry alone represented 21.6 per cent of this cost, 1.24 per cent of GDP (Safe Work Australia 2010). Compensation costs for all injuries in the construction industry amount to 0.5 per cent of the Australian construction industry's turnover, increasing building costs by up to 8.5 per cent (McKenzie and Loosemore 1997; ASCC 2009).

Numerous efforts are being made by various occupational health and safety (OHS) authorities, builders, construction unions and researchers to improve the safety performance of construction organisations and the construction industry on the whole. For example, OHS authorities in Australia have introduced more than 144 laws, 200 standards and numerous codes of practice that cover OHS in construction throughout Australia (Robinson 2002). Likewise, the Australian Federal Safety Commissioner's office has been managing an accreditation system since 2005, called the Australian Government Building and Construction Industry OHS Accreditation

Scheme, which rates builders according to their safety performance standards. This rating is notably looked at by clients in selecting builders for their projects. It is also a prequalification requirement for a builder to participate in bidding for government-funded construction projects. Equally, the Construction, Forestry, Mining and Energy Union (CFMEU) and many other WorkCover Australia registered institutions have been providing various sorts of OHS training and education to workers, supervisors and the like in the construction industry. Research centres and higher educational institutions have been producing enormous frameworks and tools for effective OHS implementation and improvement in construction. Builders have been working endlessly to make their projects safe for workers, with OHS-related resources, tools, skills, knowledge and expertise. WorkCover has been imposing heavy penalties for OHS legislative breaches.

Despite these concerted efforts by the different bodies and the proliferation of standards, frameworks, codes of practice, laws and regulations that could improve OHS in construction, the accident rate still remains at an unacceptably high level. Industry surveys and research studies suggest that the construction industry faces many unique challenges, which make OHS management and accident prevention quite complex for builders. These OHS challenges are identified below and may be attributable to construction processes, people involved and other phenomena of the industry.

Dynamic complexities of construction projects

The construction process is dynamic whereby the worksite, activities, workforce mix, materials, equipment and tools used, site layout and activity interfaces change constantly over the period of a project (Ringen *et al.* 1995; Lingard and Rowlinson 2005). The OHS risk of construction process is further heightened by increasing complexities in projects, operating and management systems, equipment specialisation and the use of potent substances. Major construction sites place a heavy emphasis on OHS management systems, which generally take the form of documented policies and procedures. However, these do not always capture hazards associated with the many non-routine jobs on constantly changing sites. Also management's urge to get paperwork and record-keeping correct for compliance purposes dilutes their attention to and emphasis on ongoing OHS problems (Wadick 2009).

Cost and schedule constraints

In a constantly highly competitive market, construction projects are often typified by tight budgetary and schedule constraints. The pressure to complete projects on budget and time tends to militate against OHS. Moreover, long work hours and the pressure for greater efficiency, the by-products of cost and time constraints, culminate in tendencies to take shortcuts, worker exhaustion, fatigue and burnout, which result in disregarding safety and encourage hazardous practices and unsafe behaviours (Hislop 1999; Charles *et al.* 2007).

Competence disparities amongst construction professionals

Effective OHS management on site relies heavily on the experience and competency level of the site management team. A considerable disparity in level of competence and experience is evident amongst construction professionals for various reasons. Employee turnover in the construction industry is relatively high, which results in a construction team where there is a mix of less experienced and experienced professionals. Additionally, due to skill migration and globalised job market systems, the construction industry in a given economy features professionals from different countries. There is a serious issue of competency and experience mismatch amongst professionals who were trained in different countries. This poses a significant challenge to effective OHS management and thus demands for the constant availability of safety training systems/programmes in construction organisations. On the other hand, it is rather difficult for construction team members to attend workshops and training owing to tight project schedules and work pressure. There are very limited opportunities and modes for learning on-the-job for construction professionals, without having to commit extra time and be away from work. Regardless, safety knowledge and competency may be acquired via self-study of OHS resources from WorkCover Australia and/ or similar OHS authorities. However, abundant OHS knowledge resides in various formats and locations, such as codes of practice, best practice manuals and databases. This is rather confusing to less experienced and less competent professionals and makes the learning cumbersome.

Migrant workers

A multicultural migrant worker demographic is one of the defining characteristics of the construction industry globally. Around 20 per cent of all workers in the Australian construction industry are overseas born and half of them are from non-English-speaking countries (DIAC 2009). Migrant workers are brought in to fill skill shortages in the domestic construction industry. However, these workers expose themselves and other workers to significantly greater risks because of their poor training, and inability to understand basic safety instructions and warning signs (Loosemore *et al.* 2009). Migrant workers come from countries where OHS standards are relaxed, have a different perception of risk and bring with them dangerous working practices from their countries of origin. Many migrants are unaware of their responsibility to manage their own and others' safety. Language barriers amongst migrant workers have direct impacts on their safety. Many migrants do not understand aspects of site induction and have difficulty engaging in toolbox talks and safety communications with their peers and supervisors. As a consequence, migrant workers represent a significant percentage in accident statistics. For example, after the UK government opened access for migrant workers in 2004, the fatality rate in the construction industry increased and 77 fatalities were recorded for the year 2006/07, the highest construction related death toll for five years (Davis and Gibb 2009).

Multilayered subcontracting system

Subcontracting has become a major feature of the construction industry. A chain of subcontractors is commonly observed in construction owing to the diversity of activities. Silberberg (1991) asserted that subcontractors are up to 90 per cent of workers in the Australian construction industry. A major concern for managing safety is the effectiveness of control over the large number of subcontractors on construction sites. Building sites with multiple subcontractors make OHS monitoring and enforcement more difficult and increase the chances of OHS non-compliance escaping undetected.

Moreover, the construction industry has comparatively low barriers to entry (Construction Training Australia 2001: 8) and many people enter with relatively low education levels (ACIL 1996: 23). Up to 60 per cent of subcontractors have no formal trade qualifications (ACIL 1996: p. xi). Conversely, OHS regulations detail several compliance requirements for subcontractors. Many subcontractors in the domestic market are unclear of these requirements and unsure what they mean. There is evidence to suggest that many subcontractors even feign compliance by using 'off the shelf' documents, without much cognitive involvement or real analysis of risks (Wadick 2005).

Constraints for management commitment

Sustained improvements in safety performance can only be achieved through collective efforts and commitments from all levels of management, including senior/top management of the organisation, site management and work supervisors. Wild (2005) argued that the poor safety performance of the Australian construction industry is significantly attributable to 'a lack of commitment on safety from senior management'. Abudayyeh *et al.* (2006) observed that top management have notable capacities and opportunities to influence and enhance the safety of the work environment. Top management commitment to safety can take two forms: (1) investing adequately in safety through reasonable budget allocations and integrating safety into organisational management settings, and (2) personally being involved in site safety activities and continual monitoring and providing feedback on site safety performance. However, there are two key impediments to this. (1) The top management in many organisations are not convinced that the investment made on safety can bring about significant financial benefits for the organisation and outweigh the expenditure. This may be due to a lack of systems for quantifying the return on safety investment in monetary terms. (2) Owing to the combined influence of the fragmented nature of construction projects and work pressure on top management, it seems rather difficult for them to be personally involved in site safety activities and monitoring on a regular basis.

These constraints are further heightened for small builders because of their limited economic/financial ability, availability of expertise and resources to deal with OHS and work culture differences (Mayhew 2002; Loosemore and Andonakis 2006). Nonetheless, small builders represent the vast majority of firms in the construction industry.

Innovation capacity drain

Continual improvements in OHS performance within an organisation rely significantly on the habit of continual learning from past experiences and feeding the insights back into practice. Chua and Goh (2004) argued that in order for the construction industry to improve its poor safety performance, it needs to learn from its mistakes and put the lessons learned to good use. And Lingard and Rowlinson (2005) asserted that a construction company may have several professionals and team players. Each professional may have some knowledge and experience in OHS. If these experiences and knowledge were collated and internalised, it could help improve the organisational learning ability and thereby OHS performance. However, most construction companies suffer from an inability to collate and leverage on past experiences and capabilities of individual members because: (1) the fragmented nature of construction projects eliminates opportunities for social learning of OHS knowledge by employees, and (2) the construction workplace culture is based on *doing*, and not in writing about doing. Innovative solutions for OHS problems are spread among site team members largely through oral stories. These stories, anecdotes and thoughts are not consolidated into a coherent body of knowledge (Wadick 2006). As a result, the valuable commodity, knowledge, that is essential for OHS innovation is lost.

Harnessing OHS with knowledge management

An analysis of the aforementioned OHS challenges suggests a pattern. The challenges may be classified into two major issues: (1) deficiencies in OHS competency, training and experience, and the inability to best utilise available OHS resources to gain the best pay-off; and (2) ineffectiveness in managing OHS affairs of dynamic and complex projects with adequate involvement of top management, site teams and subcontractors. Egbu (2008) asserted that knowledge management can play a vital role in addressing deficiencies in organisational skills and competencies. Likewise, Lingard and Rowlinson (2005) argued that the concept of organisational learning is critical to the construction industry's ability to improve its OHS performance, and suggested that, with regards to OHS, construction organisations need to develop the ability to learn. Along the same lines, Hadikusumo and Rowlinson (2004) suggested that introducing a safety knowledge management system in contractor organisations can help capture a company's collective expertise wherever it resides – in databases, on paper or in people's heads – and distribute it to wherever it can help to produce the biggest pay-off. They further added that knowledge of construction site personnel must be captured to gain advantages for establishing effective OHS systems and training programmes which can improve employees' skills. Hence, the integration of knowledge management with OHS management would be a lucrative strategy in addressing the OHS challenges indicated above and thereby improving safety performances in construction projects.

What is knowledge management?

Knowledge management (KM) is the process by which knowledge pertinent to a domain is created, acquired, communicated, shared, applied and effectively utilised in order to meet existing and emerging needs. Knowledge consists of truths, beliefs, perspectives, concepts, judgements, expectations, methodologies and know-how (Egbu 2008). Sommerville and Craig (2006) viewed KM as a systematic process of discovering, choosing, arranging, refining and presenting information in such a way that it improves an employee's comprehension (reduction in confusion, stress and competency disparity) relating to a specific area of interest. Fong (2003) regarded knowledge as the 'primary economic resource' within an organisation. Likewise, Fernie *et al.* (2003) argued that KM has the same degree of importance as labour, plant and materials within any organisation.

Forms of knowledge

Nonaka and Takeuchi (1995) categorised knowledge into two types: explicit and tacit.

Explicit knowledge:

This is the propositional knowledge of experts, which can be codified, articulated as expressions, mathematical and scientific formulas and written documents or procedures. Explicit knowledge is theory-based, scientifically constructed, easy to communicate, store and distribute and is the knowledge found in books and other visual and oral means. Examples of explicit knowledge in construction safety include accident data analysis findings and OHS management frameworks and guidelines.

Tacit knowledge:

This dimension of knowledge is know-how of everyday life. It remains tacit and cannot be codified and communicated in language or symbols. It is practice-based, context-specific, interactively derived, unwritten, unspoken and created and held internally in the brain by every normal human being, based on his or her emotions, experiences, insights, intuition and observations. For example, we can describe what we do to negotiate and control a car in a bend, but this information would not enable a learner to understand how to control a car in different bends with varying angles. They need to 'get the feel of it' by practice to understand it. In construction safety, hazard recognition is considered tacit knowledge because it relies on a site personnel's experience. Quintas (2005) and Mládková (2010) claimed that tacit knowledge may be shared (via non-codified pathways) within groups that share common learning experiences and understandings rooted in common practice; i.e. through socialisation and storytelling.

KM cycle and techniques

KM cycle consists of four key stages as illustrated in Figure 1.1 and each stage involves different techniques. Brief descriptions of each stage along with techniques used are provided below.

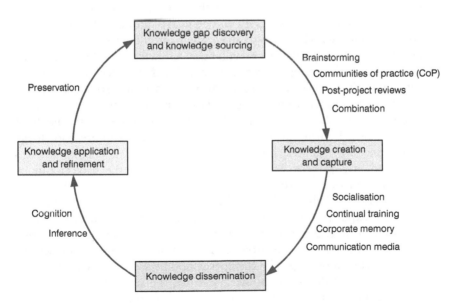

Figure 1.1 KM Cycle (adapted from Debowski 2006; Love et al. 2005)

Knowledge gap discovery and knowledge sourcing

This is the process of identifying the knowledge gap (what is known and what needs to be known) and knowledge sources related to the knowledge gap, which includes experienced staff, consultants, past cases, evidence from projects undertaken by competitors, customer/user feedback and published literature. A knowledge map/directory may be developed to identify the sources, contents and location of organisational knowledge that can be of use to address the knowledge gap (Liebowitz 2005).

Knowledge creation and capture

In this stage of KM, general principles and concepts are generated from knowledge sources to guide the construction of new knowledge. Then a conversion process takes place to refine the various ideas and principles into a specific outcome. The outcome can be either codified/explicit knowledge or embodied/tacit knowledge. Codified knowledge can be converted as models, equations and guidelines. Tacit knowledge can be shared as stories, metaphors or personal advice. The following techniques can be used to leverage knowledge creation (Debowski 2006; Al-Ghassani *et al*. 2005):

a   *Brainstorming* – This is a process involving a group of people who meet to focus on a problem and then propose as many solutions as possible through pushing the ideas to the limit. New ideas build on ideas raised by others. All the ideas are noted down and then evaluated when the brainstorming session is over to create a solution for the problem discussed.

b   *Communities of Practice (CoP)* – These are groups of people with common interests who meet to share their insights in order to develop better solutions to problems or challenges. Knowledge evolves and grows as CoP nurture sharing of concepts and ideas between members and stimulate critical reflections. These groups function through discussion lists, website forums or other form of virtual networking.

c   *Post-project reviews* – These are debriefing sessions used to highlight lessons learned during the course of a project. They capture knowledge about causes of failures, how they were addressed and best practices identified in a project. The knowledge learned in the project can be applied in subsequent projects.

d   *Combination* – This involves gathering, integrating and editing data to create explicit knowledge. For example, a construction firm collects data on lost time, productivity downturn, damage repair and other costs related to accidents from its projects. The firm subsequently processes this data to create knowledge like the correlation between safety investments and costs of accidents. In another example, the safety department of a construction firm introduces new in-house safety guidelines after studying recent accident cases and stories on construction sites of the firm as well as of its competitors.

Knowledge dissemination

Knowledge dissemination/sharing/diffusion is the spread of knowledge once it is codified or embodied. This can occur via several means as explained below (Debowski 2006; Al-Ghassani *et al.* 2005).

a *Socialisation* – This is the mode of disseminating tacit knowledge and know-how from an individual or group to other via interactions. There are different forms of socialisation, namely:
   • face-to-face meetings – a traditional approach for sharing tacit knowledge with purposeful consultations;
   • apprenticeship – apprentices work with their masters and learn craftsmanship through observation, imitation and practice; and
   • mentoring – a junior member of staff is attached to a senior member for advice related to career development, who facilitates the development of the junior by identifying training needs and aspirations.
b *Continual training* – Continual training of staff through workshops and seminars helps improve staff skills and knowledge. These can be conducted internally by senior staff or externally, where employees attend courses managed by knowledge organisations.
c *Organisational memory/corporate memory/organisational knowledge-base* – This is a system capable of storing things perceived, experienced or self-constructed beyond the duration of actual occurrence and retrieving them at a later time for application.
d *Communication media* – Media such as company newsletters, best practice documents, the intranet and visual aids such as videos, slides, charts, etc. can be used to disseminate knowledge.

Knowledge application and refinement

Knowledge creates value in use (Fahey and Prusak 1998), that is, when it is applied by organisations, it creates new capabilities and effectiveness. Thus, the utilisation of the created knowledge is a major focus of KM. Moreover, when new knowledge is applied in processes, it is tested and regularly reshaped through additional experiences and feedback. This ensures that the knowledge is reviewed and updated to reflect any new understanding and experiences; as a result, knowledge remains current. The utilisation and refinement of new knowledge takes place through cognition and inference by individuals. The refined knowledge is preserved in individuals' minds, records and business cases until it is triggered to flow through the KM cycle.

ICT for knowledge OHS

Information and communication technologies (ICT) provide the technological basis for effective knowledge management. They can facilitate knowledge creation, distribution, retention and retrieval. A range of ICT tools is available to facilitate KM as discussed below.

Modelling tools

Modelling tools provide quantitative methods to analyse subjective data to represent or acquire human knowledge with inductive logic programming or algorithms so that cognitive science and other fields could have broader platforms to implement technologies for KM development. Examples of modelling methodologies include: process modelling, cognitive modelling, pattern recognition, system dynamics, decision trees, fuzzy logic, genetic algorithm, intangible assets modelling, mathematical modelling and statistical modelling (Dekker and Hoog 2000; Hinton 2002; Kitts *et al.* 2001; Maddouri *et al.* 1998; Muller and Wiederhold 2002; Wirtz 2001; Wong 2001; Liao 2003; Kamardeen 2009).

Database technologies

A database is a collection of data organised to efficiently serve many applications by centralising the data and minimising data redundancy (McFadden *et al.* 2000). A database management system is a class of software that permits an organisation to centralise its data, manage them efficiently and provide access to the stored data for business use (Laudon and Laudon 2002).

Data mining

Data mining (DM) is a technology for knowledge discovery in large databases. It is about synthesizing useful knowledge from large collections of data by sorting through data to identify patterns and establish relationships, estimate current trends, integrate and summarise disparate information, validate models of understanding and predict missing information or future trends. DM is an interdisciplinary field that combines computer science, artificial intelligence, machine leaning, database management, data visualisation, mathematic algorithms and statistics (Liao 2003).

Knowledge-based systems/expert systems

Knowledge-based systems (KBS)/expert systems are computer programmes that emulate the decision-making behaviour of a human expert who has knowledge and experience in a particular field. They mimic human experts' ability of heuristic reasoning from the knowledge and experience gained from years of practice in solving problems. KBS/expert systems are a method of encapsulating expertise and experience pertinent to a domain in a computer 'knowledgebase', which can be interrogated by anyone via an inference engine and a search interface for decision support and/or consultation. Since

KBS/expert systems are developed by acquiring knowledge and experiences of many experts in a specific domain, they provide effective and consistent solutions for problems and produce benefit/cost results well above a human expert. Due to this high quality, KBS/expert systems have had a great commercial acceptance throughout the world. With this approach:

- valuable knowledge related to a business/process can be captured from experienced employees in an organisation and other sources, codified and stored carefully in the knowledgebase of an expert system, and
- the expert system can then be utilised by less-experienced employees for consultation and decision support. This process would allow them to add value to their existing knowledge stock in their minds.

Web technologies

Web technologies refer to tools that use internet protocols to provide platforms for exchanging data, coordinating activities, sharing information and supporting e-commerce within an organisation. Three types of web environments may be utilised to support KM within an organisation: groupware/collaborative systems, e-learning systems and web 2.0 tools.

- *Groupware/collaborative systems* are programmes that help people work in collaboration while located remotely from each other. Among the Groupware services beneficial for KM are: sharing of files, collective writing, shared database access and video conferencing.
- *E-learning systems* help the delivery of training or educational programmes via the internet, intranet/extranet (LAN/WAN), audio and videotapes, satellite broadcast, interactive TV, CD-ROM and more. E-learning can be 'on demand'. It overcomes timing, attendance and travel difficulties. Innovative enterprises have focused on e-learning in the context of knowledge management.
- *Web 2.0 tools* concentrate on socially connected web applications/services such as blogs, wikis, video-sharing sites, social-networking sites and podcasting in which people can contribute as much as they can consume.

Among the ICT tools, web technologies can play a vital role in OHS knowledge capture, codification, sharing, storage, retrieval and application for an organisation.

1 In today's knowledge economy, rapid access to knowledge is critical to the success of an organisation. Web technologies offer powerful integration and connectivity between various processes and rapid information processing abilities (Liao 2003).
2 Sheehan *et al.* (2005) argued that KM technologies used in the construction sector should establish four categories of support, including: (1) People –

supporting the profiling of people and mapping of skills, corporate yellow pages applications; (2) Projects – supporting collaborative working, document management, reviews and archiving; (3) Organisations – supporting cross-project and interdivisional working, communities of practice and idea generation; and (4) Industries – supporting external communities that unite disciplines across organisational boundaries. Inherent within these categories is the need to integrate separate knowledge areas in order to deliver added value for the organisation as a whole. Therefore the backbone of the KM technology for a construction organisation would be an intranet.

3   Sommerville and Craig (2006) asserted that islands of knowledge amongst construction supply-chain partners can become isolated, outdated or lost, thus forcing organisations to reinvent the wheel or do things they have already done. With the implementation of web technologies this issue of isolation can be overcome and the dispersed information and knowledge assets can be centrally located for ease of access and application by the construction project team. Further, they allow for easy identification of key personnel involved in a project and provide a solid foundation for the exploration of new working methods and knowledge.

4   Anumba and Ruikar (2008) claimed that online collaboration tools can facilitate easier access to information from anywhere at any time, faster transaction time, better transparency in the exchange of information, better collaboration between supply-chain partners, time and cost savings and streamlined construction business processes.

In the light of these rationales, the focus of this book in centred around the development of web-based systems for KM-enabled OHS management in construction.

KM-enabled OHS management systems for construction

Construction occupational health and safety management system (COHSMS) constitutes twelve essential elements that are to be implemented and monitored continually and holistically in any project (Lingard and Rowlinson 2005; Holt 2005; Kamardeen 2009). The twelve elements of a COHSMS are expounded below.

Declaration of policies related to OHS

Contractors should announce their OHS policies for the workplace and ensure that all employees, subcontractors and other people concerned are fully informed. An OHS policy expresses the concept that the contractor applies to improve OHS conditions of construction projects and provides for the following:

- the objective is prevention of occupational accidents;
- compliance with all OHS laws and regulations pertaining to construction; and
- proper implementation and operation of the measures stated in COHSMS.

Setting OHS targets

Contractors should set continual OHS improvement targets based on their OHS policies, taking into account: (1) past records of achievement of OHS targets and incident rates, and (2) the results of OHS investigations. The contractor should declare the goals with their stipulated period and ensure that employees, subcontractors and other people concerned understand them.

Establishing an OHS committee

In order to successfully implement the COHSMS, contractors should establish an OHS committee with the following measures:

- appoint an overall OHS controller who will oversee the implementation of COHSMS on construction sites, and stipulate his/her roles, responsibilities and authorities;
- appoint persons responsible for OHS management at every level and stipulate their roles, responsibilities and authorities;
- inform employees, subcontractors and others concerned about the aforementioned roles, responsibilities and authorities;
- ensure sufficient personnel and funds for the establishment and operation of the committee; and
- fully utilise the OHS committee for the implementation of the COHSMS.

Formulation, implementation, evaluation and improvement of project OHS plans

Contractors should formulate OHS plans to achieve their OHS targets, based on the results of investigation of project risks, and ensure that employees, subcontractors and all others concerned are informed of the plans. An OHS plan should contain:

- descriptions of actual activities and measures to be taken to reduce or eliminate their risks; and
- formats and schedules to be followed for OHS issues such as continual risk assessment, subcontractors' OHS assessment, employee consultations, OHS investigations and audits, accident investigations and OHS education/training.

Contractors should also establish procedures for implementing their OHS plans and monitoring them continuously and in an appropriate manner.

Continual assessment of risks and revision of OHS plans

Contractors should adopt specific procedures to (1) continually assess construction activity risks, and (2) determine necessary measures to be taken in order to reduce or eliminate the risks, and to meet the requirements of OHS laws and regulations. If it is found necessary, contractors should revise the OHS plan on account of changed conditions on site.

Employee consultation and feedback

In the development and improvement of OHS policies, OHS targets, OHS plans and OHS committees, contractors must set up procedures for incorporating employees' opinions and feedback.

Evaluation and monitoring of subcontractors' health and safety capabilities and performance

Contractors should establish procedures for evaluating their subcontractors' capabilities regarding OHS management, so that they can select or upgrade capable subcontractors. Contractors should also utilise the results of the aforementioned evaluation for education and training/mentoring of subcontractors.

Routine inspections and improvements

Contractors should establish and implement procedures for routine inspections and improvements of the manner in which the OHS plan is implemented.

OHS audits and revisions

Contractors should formulate a plan for periodical OHS audits and establish and implement procedures for the implementation of the audits. If it is found necessary in the audit results, contractors should improve the manner in which COHSMS is operated.

Accident investigations

Contractors should establish and implement procedures for investigating causes of incidents, should they occur, and formulating countermeasures to eliminate recurrences.

OHS education and training

Contractors should establish and implement procedures for identifying OHS training and education needs for their employees and provide them with appropriate OHS training and education.

Documentation and record-keeping

Contractors should establish procedures for documenting and handling of the following:

- OHS policies
- OHS targets
- OHS plans
- roles, responsibilities and authorities of OHS committee members
- implementation of OHS plans
- continual risk assessments
- evaluation of OHS capabilities of subcontractors
- employee consultation and feedback
- routine inspections and improvements made
- OHS audits
- accident investigations
- provision of OHS training and education

Essentially, the twelve processes above can be broadly categorised into four facets of OHS management, namely:

a OHS management system establishment
b OHS planning
c OHS training and competency improvement
d OHS implementation and monitoring

The adverse impacts of the OHS challenges that were described in the preceding section are perceptible across all four facets of OHS management process. Hence, a KM-integrated OHS management strategy is proposed below as a means for addressing these challenges.

Web-based infrastructure for KM-enabled OHS management strategy

A web-based infrastructure is proposed to assist in the practical implementation of a KM-enabled OHS management strategy while embedding the following KM functionalities within the OHS management spectrum:

1 the OHS management process should be site-boundary-independent and integrating and actively involving supply-chain members such as project teams, subcontractors, workers and top management;

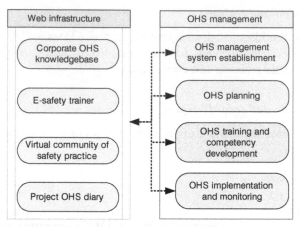

Figure 1.2 KM-enabled OHS management framework

2   OHS knowledge capture, storage and dissemination should be held as important vehicles for OHS management; and
3   learning from the OHS experiences of supply-chain members and incidents in projects and thereby continually improving the OHS system should be a normal phenomenon in the organisation.

Figure 1.2 shows the high-level conceptual model of the web-based infrastructure, which comprises four systems that can support the four facets of OHS management. The desired functional and performance descriptions of these systems are propounded below.

Corporate OHS knowledgebase

A systematically developed corporate OHS knowledgebase is crucial to continually capture and store a company's collective OHS experience and expertise wherever they reside – in databases, on papers or in people's heads – and distribute them to wherever they can help produce the biggest pay-off. Moreover, there are plenty of materials on OHS best practices from many sources and they evolve from time to time in pace with the changes in construction technology. It is imperative to keep abreast of this knowledge to better manage OHS on site. To this end, the corporate OHS knowledgebase should capture and store OHS knowledge on: (1) construction activities and their sub-steps, potential hazards associated with each sub-step and recommended precautionary measures (safe work methods); (2) hazardous substances management procedures; and (3) emergency management procedures. The corporate OHS knowledgebase can be accessed online for OHS planning. It can also facilitate learning OHS knowledge on-demand by site team members from time to time when they are about to carry out a particular construction activity or need to monitor the safety performance of a subcontractor whose work trade is different from that of the main contractor's. This can help eliminate the OHS knowledge gap in site personnel and combat challenges caused by dynamic construction processes, competency disparities, multi-level subcontracting, and tight schedule and work pressures that undermine formal workshop and training participations.

E-safety trainer

A virtual learning portal to facilitate on-demand safety training for workers on remote sites is imperative. Learning on-demand is more effective than having a standard training because on-demand leaning is activity/context-specific, and the knowledge is applied immediately and thoroughly. It is important to introduce interactive media resources in designing OHS training courses as they are easily and completely grasped by and registered in the human mind. It is also faster as opposed to reading through written descriptions. Additionally, it removes the problem posed by language barriers for less English proficient workers.

Project OHS diary

Sustained improvements in safety performances on sites may not be possible without an established safety culture at the organisational level. Safety climate on construction sites is a product of safety culture in the organisation. Safety culture is concerned with the ability of an organisation to manage safety with a top–down approach. This is heavily underpinned by the ability of the top management to analyse project-based OHS data and develop innovative strategies from time to time to overcome OHS challenges. Holding OHS data in papers on sites makes it difficult to capture, store, retrieve and analyse it. It is therefore essential to maintain a centralised online OHS diary that would allow recording of day-to-day OHS implementation data on segregated sites. This data can serve three critical purposes for top safety management, including: (1) tracking OHS actions on site, measuring performances and triggering control or corrective measures, if required; (2) demonstrating OHS compliance; and (3) deriving innovative strategies, policies and systems for continually improving safety in current and future projects.

Virtual Community of Safety Practice

The formation of a Community of Safety Practice (CoSP) is necessary to integrate professionals from a range of organisations such as contractors, subcontractors, suppliers, manufacturers, OHS consultants and other OHS stakeholders. A homepage for the CoSP that recognises the existence of it is vital for encouraging safety experts to subscribe to the community as members. This can also function as a safety expert yellow pages in which experts are mapped on their professional and specific trades and experiences.

Users can access this service to locate relevant experts and get help in solving OHS problems in their projects. Additionally, a discussion portal is vital to foster threaded discussions by the CoSP members surrounding concerning OHS issues. The contents of threaded discussions should be preserved in a knowledgebase for future use, which will avoid the possibilities of initiating discussion sessions for already solved problems. The web portal of CoSP could also provide an opportunity to organise industry-wide OHS symposiums and workshops for construction professionals from time to time. These events would be effective platforms for OHS knowledge dissemination and learning, and the facilitation of such events would be easy with the CoSP web portal.

Structure of the book

The web-based infrastructure strategy discussed above drives the theme of the book and sets the foundations and the systems in the infrastructure form individual chapters. To this end, the book consists of six chapters, and the contents of each chapter are outlined below.

- This first chapter has placed the theme of the book in perspective and laid the foundation for discussion in other chapters.
- Chapter 2 elaborates on the development methodology of a corporate OHS knowledgebase to assist with OHS planning.
- Chapter 3 discusses the development of a web-based interactive system for affective OHS training to workers that could influence workers' attitudes towards safety and thereby improving safe behaviours.
- Chapter 4 describes the development of an online project OHS diary that facilitates effective and real-time record-keeping, tracking and monitoring of OHS implementation on site. Emphasis is placed on how the diary can be utilised (1) to improve OHS performance continually based on lessons learned from the site; and (2) to demonstrate OHS regulation compliances in construction projects.
- Chapter 5 reports on the development of a virtual community of safety practice to promote social learning of safety and cultivate a strong safety culture within the construction industry and thereby improving overall safety performance.
- Chapter 6 brings together all of the subthemes and findings in previous chapters into a coherent model and reports its integrated applications in the construction industry. The chapter also discusses key considerations for practical applications.

Discussions in any chapter of the book have been laid out independently of other chapters such that the chapters can be read in any sequence without compromising readers' comprehension of the contents. Nevertheless, it is important that readers appreciate the connection of the chapters to the facets of OHS management.

2 Corporate knowledgebase for OHS planning in construction

Introduction

Construction accidents cause many human tragedies, demotivate construction workers, disrupt construction processes, delay progress and adversely affect the cost, productivity and reputation of the constructor (Kartam 1997). Pre-project and pre-task OHS planning is among the critical measures required to achieve a zero accident target (Saurin et al. 2004). The ability to identify health and safety hazards as early as possible and implement adequate controls is vital to a project of any size and scale (Cheung et al. 2004). The deployment of effective OHS planning and control techniques on construction sites to prevent accidents can therefore have significant human, social and financial impacts. Moreover, OHS planning often appears as a core requirement in OHS regulations and standards. For instance, OHS regulation 2001 of New South Wales, Australia, states that: 'The principal contractor for the construction work must ensure that a site specific OHS plan is prepared for each place of work at which the construction work is to be carried out before the work commences' (WorkCover NSW 2001a).

Chua and Goh (2004) and Saurin et al. (2004) define OHS planning as a process which consists of three interdependent components: (1) establishing the necessary tasks to be undertaken, (2) identifying the existing hazards and potential risks and (3) defining how the risks will be controlled. A hazard is simply a situation that has the potential to harm people physically or psychologically. Fatalities, injuries, health damages or ergonomic problems are the possible outcomes of a person being exposed to hazards. The project team is required to identify task-based hazards and potential risks by analysing: (1) the nature of the task and sub-tasks or job procedures, (2) location of the task, (3) interface with other tasks, (4) materials, equipment and tools used and (5) nature of the work crew (language barriers, worker mix and demography) (Toohey et al. 2005). Having identified hazards and potential risks in a task, the project team is required to implement: (1) safe construction practices, (2) personal protective equipment use, (3) permit-to-work systems and (4) housekeeping systems (Holt 2005; Teo et al. 2005; Kamardeen 2009). Thus, the project team is required to be well-versed in the knowledge and skills related to (1) the identification of hazards and potential risks in each job procedure in a task and (2) the implementation of risk control measures for job procedures. Nonetheless, OHS planning faces significant difficulties and challenges for the following reasons.

• Construction projects have large scopes of work and every project is unique. Construction processes on a given site are dynamic. Factors such as the working environment, activity, workforce mix, equipment and tools usage and site layout change rapidly and constantly over the period of construction. This nature of construction demands substantial OHS knowledge from site personnel. However, it is unlikely that a site team possesses all the knowledge and experiences required to identify every potential hazard in the broad scope of work, and develop an effective OHS plan (Carter and Smith 2006).

• OHS knowledge is in abundance and it resides in various formats such as codes of practice, best practice manuals and databases. For example, Robinson (2002) highlighted that there are more than 144 laws, 200 standards and numerous codes of practice that cover OHS in the construction industry throughout Australia. Additionally, these knowledge contents evolve over time. It is (1) unlikely that project teams would know the whole contents of these resources for developing effective OHS plans and (2) nearly impossible for them to refer to these resources, due to their tight project schedules and work pressure.

• OHS planning relies heavily on the experience and competency of the site team. The process of identifying hazards, assigning appropriate levels of risk and selecting the most efficient control requires extensive field knowledge and experience (Chua and Goh 2004). Nevertheless, leveraging on the experience of different site teams is hampered by the fragmented nature of construction sites. Carter and Smith (2006) reinforced this claim that information required to improve hazard identification and OHS planning already exists, albeit in a fragmented state. The individual creating OHS plans is not usually able to obtain all the necessary information due to the lack of a standardised approach to capturing and sharing OHS knowledge across projects.

• In the globalised economic system, construction industries feature professionals from different countries. The serious issue of skills shortage is a significant contributor to ineffective OHS planning (Trajkovski and Loosemore 2005).

Carter and Smith (2006) suggested that implementing the concept of knowledge management (KM) into OHS planning will eliminate these challenges and improve the OHS planning process continually. KM is a method of exploiting or transforming knowledge from various sources as an asset for organisational use to facilitate continuous improvement. It is a recent and evolving practice for construction organisations. KM can help capture a company's collective expertise wherever it resides – in databases, on paper or in people's heads – and distribute it to wherever it can help produce the biggest pay-off (Hadikusumo and Rowlinson 2004). Lingard and Rowlinson (2005) also argued that the concept of KM and organisational learning is one that critical to the construction industry's ability to improve its OHS performance. Lehtinen *et al.* (2005) advocated that KM is needed in OHS so that: (1) a company's scarce resources are not wasted by duplicating work, (2) OHS information is easily available and accessible to all users of that information and (3) information is kept as a structured entity instead of as fragmentary and unorganised bits. Researchers have suggested different approaches to implement KM and organisational learning into OHS planning, as expounded below.

- Sorine and Walls (1996) urged the need for a system that facilitates the dynamic compilation, updating and dissemination of: (1) information on standard job procedures, which includes steps that the employee performs, potential hazards of these steps and recommended controls, (2) regulatory mandates and (3) related safety information for OHS planning.
- Lehtinen *et al.* (2005) suggested that organisations identify knowledge and information gaps in OHS and draw up a plan for producing relevant packages of OHS information and disseminating to projects through the internet.
- Carter and Smith (2006) explored various options to develop an OHS knowledgebase and concluded that a web-based tool would be a better approach as it would be available on virtually every construction project within an organisation. Moreover, it would be platform independent and could be used within an ordinary web browser without the need for high-specification hardware and software.

It was therefore hypothesised that a centrally located, dynamic virtual OHS knowledgebase that is updated continually would help address the challenges faced in effective OHS planning. The knowledgebase should contain: (1) breakdowns of construction tasks into job procedures, and hazards and potential risks in each job procedure, and (2) regulatory mandates, best construction practices and risk control measures for the job procedures. Hence, the research described in this chapter aimed at developing an effective OHS planning system that simultaneously implements KM and organisational learning into OHS. The aim was achieved via a systematic three-step approach.

1   Developing the concepts and contents of a knowledge-based OHS planning framework.
2   Formulating the conceptual model of a web-based system for knowledge-based OHS planning.
3   Prototyping and validating the web-based system.

Construction features large scopes of work and multitudes of trades. It was nearly impossible to cover all the trades in a construction project in this research due to time constraints. Hence, the study was demonstrated in the context of conventional formwork for suspended slabs. Nonetheless, the knowledgebase can be populated with information for other trades using the frameworks and the system model incorporated into the proposed system. Additionally, the study was carried out in the context of the New South Wales construction industry, Australia. The codes of practice and other regulatory mandates that are discussed in this study are specific to the said industry context.

The study was carried out in four stages: model development, knowledge acquisition, system implementation and system validation. First, a thorough literature review supported the identification of terms of reference for the proposed system and consequently the development of the conceptual model of the system. Subsequently, the knowledge acquisition to populate the knowledgebase of the system was performed by analyses of documents and publications that describe job procedures, hazards, best practices and control measures. Then the conceptual model was implemented using Joomla!, an open source contents management system. Finally, the system was validated against the research hypothesis with construction industry practitioners. This chapter discusses the content in a slightly different order than the actual study steps. First, the context for the study is laid out, followed by the knowledge acquisition for the knowledgebase, then the conceptual model of the proposed system is elaborated, followed by system implementation and validation, and finally conclusions are discussed.

Acquisition of OHS planning knowledge

The acquisition of OHS planning knowledge for slab formwork had two folds: documents analysis and practitioners reviews. In the documents analysis phase, codes of practice, best practice manuals, textbooks and research publications were analysed thoroughly. These analyses helped identify job procedures, hazards, risks, causes and best practices for the job procedures. Subsequently, an OHS profile was developed to contain the knowledge. In the next phase, the OHS profile was subject to review and critique by two experienced project managers from two different construction organisations. Following that the OHS profile was improved. The findings of the knowledge acquisition exercise are described below in detail.

Formwork job procedures and hazards

Conventional slab formwork is made of plywood or metal sheathing for decking. Sheathing is supported by horizontal members called joists or runners. Joists are supported by another set of horizontal members perpendicular to the joists, called stringers or bearers. The stringers are supported by vertical members called shores. Vertical timber shores can be replaced by the scaffold type, which has been proven to be more efficient because of its high number of reuses and its height, which means that no splicing is typically required. The scaffold-type shoring system consists of two vertical steel posts with horizontal pipes between them at regular intervals. Adjustable screw jacks are fitted into the steel posts at both ends. The top jacks are fitted into steel caps called T heads. The bottom jacks are fastened into rectangular steel plates. Adjacent vertical steel posts are braced together by steel X braces.

Job procedures involved in erecting formwork for slabs and their hazards were identified from CSAO (2009), WorkCover NSW (2001b), Spielholz *et al.* (1998), Jurewicz (1988) and Borden (1986). There are nine job procedures in this task and the hazards and risks prevalent in these job procedures are described below.

Setting out

The first procedure involved in formwork is to set out form arrangements to the requirements of plans and specifications. Hazards associated with this procedure are: (1) slippery conditions due to rain, snow or ice causing trips and falls, and (2) exposure to ultraviolet light and glare causing skin cancer, sunburn and damage to eyes.

Moving/lifting materials from stacked locations to the work area

Moving materials to work areas may involve the use of cranes for hoisting from lower levels, a forklift or manual movements. Hazards inherent in this job procedure comprise the following. (1) Manual handling involves repetitive lifting and carrying of materials. These cause strains, sprains, back injuries and falls by tripping. Plywood and other sheet materials can be difficult to handle in the wind. The added and unexpected extra load imposed by a sudden gust of wind can throw a worker off-balance and lead to a fall. (2) Use of a forklift may cause struck-by accidents due to driver's inexperience, reversing buzzer or light not working, the driver or workers not watching and/or defined areas for vehicle operation not being marked. There are also risks of toppling of forklift due to improper loading and operation of the forklift. (3) Using a crane to hoist materials from lower levels to the working area can cause strike-by injuries and falls when unloading materials.

Preparing foundation/ground and placing sole plates

If the foundation for the formwork is unstable because of soil condition or the suitability of sole plates, it can lead to falls from unstable formwork or the collapse of formwork.

Erecting shores/props/vertical supports and bracing

Shoring for formwork is provided by joining individual tubular scaffold members together to reach the required ceiling height. The erection involves many sub-procedures: initial set-up of first-level frames and bracing, placing working planks on first frames, erecting second-level frames and bracing, transferring working planks from first-level frames to second-level frames, and erecting third-level frames and bracing. These sub-procedures can be broadly categorised as: (1) erecting frames and bracing, and (2) placing working planks.

Hazards present in erecting frames and bracing include the following. (1) Manual handling, involving lifting and passing frames up, causes strains, sprains and back injuries. (2) Working at heights/on ladders to fix frames or bracing leads to falls due to loss of balance. (3) Incorrectly secured bracing and frames jam fingers. (4) Passing up tools and equipment causes hits by falling tools and equipment. Hazards inherent in placing and using working planks are: (1) manual handling to lift planks up onto frames lead to strains, sprains and back injuries, (2) use of ladders causes falls and (3) unsecured/faulty/wrong/unstable/slippery planks can cause falls from working planks.

Placing bearers (stringers) and joists (ledgers)

Hazards associated with erecting bearers and joists are the following. (1) Manual handling, involving lifting, manoeuvring and passing up bearers and joists, causes strains, sprain, and back injuries. (2) Working at heights to fix bearers and joists in place can lead to falls from frames/working platforms/ladders. (3) Use of tools such as hammers, pry bars and wedges, and materials such as joists and bearers can cause struck-by injuries.

Placing form ply deck

Placing form ply deck involves placing sheathing and spraying form release coating. Among the hazards in this job procedure are the following. (1) Manual handling of plywood causes strains, sprains and back injuries. (2) Working at heights on deck leads to falls through penetrations in the deck or from lead edges. These are caused by the loss of balance due to gusts of wind and/or unguarded penetrations and edges. (3) Slippery decks due to wetness, oil, sawdust or other objects left on the deck cause people to slip and fall over objects. (4) The use of rotating power saw blade causes serious cuts, hearing damage due to noise and electrocution due to a faulty power saw or unprotected extension lead. (5) When cutting plywood, there is exposure to wood dust and epoxies that, when inhaled, can cause asthma and bronchitis. (6) Direct contact with form release coating can cause skin rashes and eye irritation.

Mounting intermediate props for the deck from below

To provide further support to the formwork, intermediate Acrow props are introduced for the deck from below. This job procedure risks: (1) strains, sprains and back injuries when lifting and manoeuvring props (manual handling), and (2) inadequately secured props falling on workers.

Stripping formwork

Formwork stripping hazards include the following. (1) Falling materials hit workers below. (2) Materials, tools and sharp objects underfoot cause people to trip over and puncture wounds. (3) Manual handling of heavy or awkward forms, panels and other components leads to strains, sprains, back injuries, cuts and crushed fingers. (4) Prying forms loose from concrete present risks of overexertion, lost balance, and slips and falls. Particles of wood and concrete can cause eye injuries. (5) Insufficient lighting can cause workers to walk into objects, slip, trip, fall and sustain other injuries. (6) Poor access and inadequate working space because of stripped materials not progressively removed cause trips over materials. (7) Working at heights on platforms and ladders to strip formwork may cause falls.

Repair formwork for reuse

Repair of formwork for reuse involves the following: (1) timber components are de-nailed, cleaned and stored/stacked safely for reuse or removal from site, (2) steel components are cleaned, oiled and stored/stacked to manufacturers' maintenance recommendations, and (3) damaged formwork components are safely discarded after stripping. Hazards in repairing formwork for reuse include the following. (1) Dust and flying particles during the cleaning of formwork can injure eyes. (2) Handling of tools and equipment to clean and repair formwork can lead to cuts, struck-by injuries and crushed fingers. (3) Materials handling and protruding formwork parts can cause sprains, strains and back injuries. (4) Debris and sharp objects can cause foot injuries, trips and falls.

Developing an OHS profile for formwork

Following the exploration of hazards and risks in job procedures for slab formwork, the documents analysis was directed to identify best practices for these operations. Codes of practice of OHS authorities in Australia and other published literatures were analysed meticulously. These include *Formwork Code of Practice 1998* (WorkCover NSW 1998), *Formwork Code of Practice 2006* (Queensland Government 2006) and other work by CSAO (2009), IAPA (2008), Department of Commerce (2007), MMA (2005), Hanna (1998) and Sommers (1981). Subsequently, an OHS profile for slab formwork was developed as illustrated in Table 2.1. The OHS profile identifies hazards, risks and their root causes in each job procedure in slab formwork, and then the best practices for each job procedure that avert risks are suggested.

Knowledge-based OHS planning framework

A framework was developed for the knowledge-based OHS planning strategy advocated in this study as depicted in Figure 2.1. The framework explains the activity flow and resources essentially involved in the planning process.

- The kernel of knowledge-based OHS planning is the OHS knowledgebase, which is constructed by collating knowledge from codes of practice, best practice manuals and research publications. This knowledgebase is updated continually by a dedicated knowledge worker when new knowledge emerges.
- Knowledge-based OHS planning involves analysing the situational variables, retrieving OHS knowledge from the knowledgebase, devising effective plans for the situation, implementing them on site and then updating the knowledge base with new knowledge from the site.
- Situational variables include the type of activity to be carried out, work method, location, interacting work, materials used, plant used and the nature of operatives. These variables are studied meticulously for each activity, and relevant potential hazards and control measures are tapped from the knowledgebase. These lead to the development of an OHS plan for the activity.
- The OHS plan is implemented on site and monitored. If any new hazards and risk controls are identified, these will update the knowledgebase.

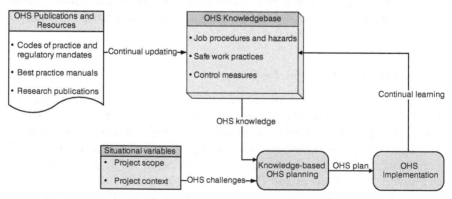

Figure 2.1 Knowledge-based OHS planning framework

Table 2.1 OHS profile for slab formwork

---

1. Setting out

Hazards	*Risks*	*Causes*
Slippery ground/ floor condition	Slip, trip and fall on the floor	Rain/snow/ice Insufficient ground/floor preparation
Working at heights	Falls through unprotected upper floor edges and openings	Unprotected upper floor edges and openings
Exposure to UV light and glare	Skin cancer, sunburn and damage to eyes	Ignoring/not providing personal protective clothing (sunscreen 15+, shirt and flap on helmet)

Best practices
1. Ensure that appropriate fall protection systems (perimeter screens/scaffold, guardrails or barriers) are in place around the floor before working at heights.
2. Cover or guard all the openings on the floor.
3. Provide suitable and safe access to and from the worksite.
4. Ensure the work area is even and free from rubbish and slippery agents such as water, grease/oil, snow, etc.
5. Provide appropriate PPE to protect workers from UV ray and glare (sunscreen 15+, shirt, flap on helmet, sunglasses, etc.).
6. Clearly mark the formwork area or provide barriers to prevent formwork setting out operatives from entering other areas of hazards.
7. Introduce physical barriers and signboards to prevent unauthorised persons from entering the formwork area.

---

2. Moving materials from stacked locations to the work area

Hazards	*Risks*	*Causes*
Manual handling of formwork materials	Strains, sprains and back injuries Trips and falls	Improper stocking of formwork materials causing additional handling to turnover and lift. Repetitive lifting and carrying Gust of wind

Best practices
1. Establish clear access/egress to allow for the safe movement of materials around the site.
2. Material storage/lay down areas should have enough space and lighting.
3. The ground conditions of storage/lay down areas and access/egress should be even, firm and unslippery.
4. Wherever practicable use mechanical aids such as carts, hand trucks and dollies to move and place large and heavy loads.
5. Have materials placed at the working level and readily accessible to the leading edge.
6. Always ensure components are de-nailed before handling, and wear adequate PPE (e.g. gloves, helmets, sun glasses, boots, etc.).
7. Take precautions such as: (1) test the weight of the load to ensure it can be lifted securely, (2) grip the load securely, (3) protect hands against pinch points and struck-by injuries, (4) practise good team lifting, (5) get help with awkward loads, (6) always use mechanical devices and aid provided and (7) do not rush.
8. When pushing and pulling are involved: (1) push rather than pull, (2) avoid overloading, (3) ensure the load does not block vision and (4) never push one load and pull another at the same time.
9. Ensure the work areas are continuously tidy and free of obstructions that may prevent safe movement of materials and people.
10. To avoid injury to muscles, ligaments and other soft tissues do warm-up/stretching exercises at the beginning of the day and cool down/stretching exercises at the end of the workday.

Hazards	Risks	Causes
Use of a forklift	Tip over	Improper loading, speed and
	Workers struck by forklift and	turning techniques
	falling loads	Inoperative reversing buzzer,
	Collision with fixed objects	light or brake
	Roll away forklift	Inattentive operator or workers
	Driving off loading floor	No defined areas for plant
		operation

Best practices
1.   Complete the pre-shift inspection to make sure brake, backup alarm, horn, seatbelt and light are operational.
2.   Wear the seatbelt.
3.   Wear proper PPE (hearing protection and safety glasses).
4.   Observe safe operating speeds for conditions.
5.   Sound horn when approaching doorways, aisles, pedestrian walkways, pedestrian doors and blind corners.
6.   Use smooth and safe turning techniques.
7.   Travel in reverse if carrying a load that obstructs forward vision.
8.   Observe load handling/stacking rules.
9.   Keep the load high enough to avoid fetching up on inclines or uneven surfaces. Mast (tilt) the load back as soon as possible after picking up.
10. No riders.
11. Always be watchful of the work path and the surrounding area, and while backing up.
12. The parking brake must be set before the operator leaves. If the operator is more than 7.5 m away from the forklift or out of direct sight of it, the engine must be shut off.

Hazards	Risks	Causes
Use of a crane to hoist formwork	Falls from ladders or deck when	Unstable deck or ladders
materials	slinging loads	Loss of balance due to force
	Struck-by injuries	from the load
		Sling is not secured properly

Best practices
1.   All lifting gear such as slings, hooks, shackles, material boxes, straps and lugs should be inspected for damage and wear before lifting.
2.   Lifting slings should be used when lifting formwork frames. Loads of joists, bearers and form ply should be strapped together and lifted in a flat position.
3.   Tag lines should be used to guide and control suspended loads. And areas beneath suspended loads should be clear of persons.
4.   Areas in the vicinity of materials or loads being moved should be clear of persons when moving long materials such as joists, bearers, planks and frames to prevent striking persons nearby.
5.   Use a platform to sling materials. The platform should be at least 450 mm wide and have edge protections. There should be a safe means of access to the platform for persons to access the platform. Ladders may be used and they should be secured at the top to prevent movement.

3. Preparing foundations/ground and placing sole plates (sills/mudsills)

Hazards	Risks	Causes
Unstable foundation for	Falls from unstable formwork	Soft or uneven ground – under
formwork	Collapse of formwork	preparation
		Packing under the sole plate
		washed/blown away
		Unsuitable sole plates – under
		strength

Best practices
1.   Soil should be firmly compacted under sills and proper drainage to be provided to prevent ponding of water in the area.
2.   Soil, if unstable, removed and replaced with stabilised materials under the sill; mudsills not supported on frozen ground.
3.   Check the suitability of mudsill sizes for shore loads and bearing capacity of soil.

4. Complete the ground level slab wherever possible before shoring is erected for suspended slabs.
5. The base of the formwork must be stable and even to prevent the risk of collapse.
6. Safety features of power tools used must be checked before use.
7. Maintain continuous good housekeeping to keep work areas and passages safe and unslippery.
8. All persons involved in erecting sole plates should wear appropriate PPE for the nature of the site and sun protections.

4. Erecting shores/props/vertical supports and bracing

Hazards	Risks	Causes
Manual handling of scaffold	Strains, sprains and back injuries	Lifting and passing frames and working planks up
Standing on planks/ ladders to work and/or lift frames/ planks	Falls from working planks/ ladders	Loss of balance when lifting frames and planks Collapse of working platform/ ladders Unsecured/faulty/unstable working planks/platforms and ladders
Incorrectly secured frames and bracing	Jammed fingers	Pins not in position Damaged/rusty spigots
Passing up tools and falling tools from working platforms	Hit by falling tools	Poor coordination between two workers Loss of grip Not using the bucket/tool box to pass tools up

Best practices
1. When lifting and passing frames up, take such precautions as: (1) test the weight of the frame to ensure it can be lifted securely, (2) grip the frame securely, (3) use gloves to protect hands from struck-by injuries and pinching, (4) practise good team lifting and proper coordination between team members when using ladders/working platforms to avoid loss of balance, (5) get help with awkward loads, (6) wherever possible use mechanical devices and aid provided, and (7) do not rush.
2. Ensure access ladders and other working platforms are safe and well-secured before climbing on. Working platforms must be at least 450 mm wide, cleats should be used to prevent planks from slipping off the frames, and guardrails should be fixed if the height of the platform is greater than 2 m.
3. Make sure adequate bracing, vertical support and continuity of support are available, and no member is defective, corroded or deformed.
4. Frames must be erected in a progressive manner to ensure both installers' safety and the stability of the structure.
5. Braces must be attached to the frames as soon as practical. Make sure pins and spigots are in right positions and well-secured.
6. Install edge protections on the frames as they are erected to prevent the risk of falls on edges of formwork frames.
7. Use a bucket to pass tools up and keep all the tools at heights in a container.
8. Safety features of power tools used must be checked before use.
9. Maintain continuous good housekeeping to keep work areas and passages safe and unslippery.
10. All persons involved in erecting shoring and bracing should wear appropriate PPE and sun protections.
11. Only persons involved in the construction of the formwork should be located in the formwork construction zone. Any person not involved in the erection of shoring and bracing should be excluded.

5. Placing bearers (stringers) and joists (ledgers)

Hazards	Risks	Causes
Manual handling	Strains, sprains and back injuries	Lifting, manoeuvre and passing up bearers and joists Loss of grip on bearers/joists Timber/s too heavy
Working at heights/on ladders	Falls from working platform/ladders/fixed joists	Loss of balance when lifting and positioning bearers and joists Unsecured joists roll when walked on Slip when climbing
Use of tools such as hammers, pry bars and wedges and handling members	Struck-by injuries	Inattentive use of tools Fall of tools from above

Best practices
1. When lifting and passing bearers and joists up, take such precautions as: (1) test the weight and length of the member to ensure it can be lifted securely, (2) grip the member securely, (3) use gloves to protect hands from struck-by injuries and splinters, (4) practise good team lifting, (5) get help with awkward loads, (6) always use mechanical devices and aid provided, and (7) do not rush.
2. Do not stand/walk on bearers to lift and place other bearers or joists. Use an intermediate working platform instead.
3. Working platforms/intermediate platforms for lifting bearers up should not be more than 2 m high from the floor. The height between the intermediate platform and the top of the frame should not be more than 2 m, working platforms must be at least 450 mm wide and cleats should be used to prevent planks from slipping off the frames.
4. Practise team work for receiving and spreading bearers and joists. This will avoid falls due to loss of balance.
5. When only single bearers are placed in the U-head, the bearer must be placed centrally in the U-head, with minimal cantilevers.
6. Where the top of the supporting frame consists of a flat plate, the bearer must be nailed or otherwise effectively secured.
7. Joist should spread on bearers from a platform located within 2 m of that surface, underneath the deck being constructed.
8. Place joists on the bearer in a progressive manner from the work platform located directly below the area to be worked on, and spaced at 450 mm centres (maximum) or so that the gap between joists does not exceed 400 mm.
9. Nail the joists to bearers or otherwise effectively secure them. Deep joists need to be laterally braced to prevent overturning.
10. Use buckets to pass tools and other objects to workers above. Tools (hammers, pry bars, nails, wedges etc) used for the operation should be kept in a container.
11. Cantilevered bearers and joists must be secured against uplift prior to loads being supported by them.
12. All persons involved in the erection of bearers and joists should wear PPE, sun protections and appropriate gloves.
13. Only persons involved in the construction of the formwork should be located in the formwork construction zone. Any person not involved in the erection of bearers and joists should be excluded.

6. Placing form ply deck

Hazards	Risks	Causes
Manual handling	Strains, sprains and back injuries	Lifting up plywood – sheets too large/additional load by wind force
Working at heights on decks	Falls through penetrations in the deck/from lead edges	Unprotected penetration in the deck Unprotected lead edges
Slippery decks	Slip and fall over objects	Wetness, oil, saw dust or other objects left on the deck
Cutting plywood using rotating power saw	Serious cuts Electrocution Hearing damage Asthma and bronchitis from saw dust Eye damages from plywood splinters/flying objects from the saw	Faulty power saw/blade unguarded Unprotected extension lead Earth leakage switch not installed/faulty No/incorrect PPE used for the task
Contact with form release coating	Skin rashes Eye irritation	No/improper PPE used for the task

Best practices
1. When lifting and passing plywood up, take such precautions as: (1) test the weight and size of the plywood to ensure it can be lifted securely, (2) grip the plywood securely, (3) use gloves to protect hands from struck-by injuries and splinters, (4) practise good team lifting and proper coordination between team members when using ladders/working platforms to avoid loss of balance, (5) get help with awkward loads, (6) wherever possible use mechanical devices and aid provided, and (7) do not rush.
2. Leading edge and perimeter protections must be provided on edges where the potential fall distance is 2 m or more. The most effective means of providing edge protection is by providing perimeter scaffolding.
3. Warning signs should be displayed (e.g. Danger, Keep clear of the edge).
4. A formwork deck must be laid in a progressive way such that persons will be provided with a method of preventing them from falling below the deck.
5. Form ply should be placed on the joists with the installer located behind the sheet as it is positioned whilst standing on the previously laid sheet or provided platform.
6. Installers should not walk on joists for placing form ply.
7. Nail or otherwise secure form ply to the joists as soon as practical.
8. Cover or protect all penetrations. Covers must be securely fixed and clearly signed to indicate they are protecting a penetration (e.g. Danger Hole Under).
9. The leading edge should be free of oil, sawdust and obstructions to reduce the likelihood of slips and trips.
10. Use a bucket to pass tools up and keep all the tools at heights in a container.
11. Safety features of power tools used must be checked before use and extension leads must be protected.
12. Maintain continuous good housekeeping to keep work areas and access ways safe and unslippery.
13. All persons involved in placing form ply deck should wear appropriate PPE and sun protections. PPE should be against hearing damages, asthma and bronchitis, eye damages and skin rashes.
14. Only persons involved in the construction of the formwork should be located in the formwork construction zone. Any person not involved in construction of the formwork deck and support structure should be excluded.

7. Mounting intermediate props for the deck from below

Hazards	*Risks*	*Causes*
Inadequately secured props	Acrow props fall on workers	Prop not secured properly Prop not laced/tied in
Manual handling	Strains, sprains and back injuries	Lifting and manoeuvring props

Best practices
1. Take precautions such as: (1) test the weight of the prop to ensure it can be lifted securely, (2) grip the prop securely, (3) use gloves to protect hands against pinch points, (4) practise good team lifting, (5) get help with awkward loads, (6) always use mechanical devices and aid provided, and (7) do not rush.
2. Check the quality of props, pins and sole plates before use. Any damaged, deformed, or worn (due to run-out, corrosion or rotting) items should be rejected.
3. Individual shores should be laced both ways with continuous runners, and braced laterally.
4. Ensure pins are in correct positions and tight.
5. Whenever sole plates are used, ensure they are firm and strong to support the load.
6. Wear appropriate PPE.

8. Stripping formwork

Hazards	*Risks*	*Causes*
Stripping materials from below	Hit by falling materials and tools	Area not barricaded off Uncontrolled lowering of deck and materials
Prying forms	Slip and fall	Overexertion and loss of balance
Working at heights on platforms/ ladders	Fall from formwork frames, working platforms/ladders	Collapse of frames/working platforms/ladders when dismantling due to loss of balance or bracing Slip off the working platform/ ladder
Manual handling	Strains, sprains and back injuries	Lifting and passing down form materials Handling heavy/awkward forms, and panels
Insufficient lighting	Walk into objects, slip, trip, fall and sustain other injuries	Poor lighting provided to the work face when stripping
Materials, tools and sharp objects underfoot	Trip and fall Puncture wounds	Stripped materials not progressively cleaned up No/incorrect PPE for the task
Poor access and inadequate working space	Trip over materials	Timbers and other materials stacked across access ways Stripped materials not progressively cleaned up

Best practices
1. Cordon the stripping area off and display signs to keep other people out of the area (e.g. Danger – formwork stripping in progress – authorised persons only).
2. Provide adequate lighting for the work area, if natural lighting is inadequate.
3. Before stripping formwork, ensure penetrations in the slab that will be exposed as the formwork is stripped are covered and protected prior to the commencement of the stripping operation.
4. Only remove the formwork support system (props or frame) from under the sheet to be removed. Once the sheet is removed, progressively continue removing adjacent sheets.

5. The person levering off the sheet should not stand directly under the sheet but stand far enough away on a working platform and use a pinch-bar to lever the sheet off. The working platform must be at least 450 mm wide and cleats should be used to prevent planks from slipping off the frames.
6. The personnel assigned to stack removed frames and sheets should stand clear of the removal face while the other person is levering off the sheets.
7. All nails should be removed from the formwork material during dismantling. High tensile nails should be removed with an appropriate tool to prevent nails becoming projectiles when being removed.
8. Keep the removed sheets and frames in separate stacks, away from access ways, as soon as they are removed.
9. Progressively remove the stripped materials and any other debris from the floor.
10. All workers involved in stripping slab formwork must wear such PPE as safety helmets, eye protections, safety gloves and foot wears, and appropriate and comfortable safety harnesses.
11. In stacking and removing frames and sheets:
    1. Perform warm-ups before work to prevent sprains and strains when lifting and manoeuvring.
    2. Arrange for team lifting for stacking frames and sheets when it is impractical to use mechanical assistance. Whenever necessary use dollies, carts or other mechanical devices.
    3. Take job rotations/adequate rest breaks when carrying materials and for work that involves repetitive bending, twisting, over reaching or work overhead.
    4. Provide mechanical aids to remove stacked frames and sheets from the work area. Lifting equipment could include cranes, forklifts, or electric pallet trucks/stackers.
12. Safety features of power tools used must be checked before use and extension leads must be protected.
13. Only persons involved in the construction of the formwork should be located in the formwork construction zone. Any person not involved in the stripping of formwork should be excluded.

9. Repair formwork for reuse

Hazards	Risks	Causes
Dust and flying particles when cleaning and repairing formwork	Eye injuries	No PPE used
Handling of tools for cleaning and repairing formwork	Cuts Struck-by injuries	In attentive use of tools No PPE used
Materials handling	Sprains, strains and back injuries Crushed fingers	Lifting and carrying heavy and bulky materials Awkward postures
Debris at workspace	Puncture wounds Trips and falls	Debris not progressively cleaned up from the work area

Best practices
1. When lifting and moving stripped formwork materials, take such precautions as: (1) test the weight and size of the material to ensure it can be lifted securely, (2) grip the material securely, (3) use gloves to protect hands from struck-by injuries and splinters, (4) practise good team lifting, (5) get help with awkward loads, (6) always use mechanical devices and aid provided, and (7) do not rush.
2. Provide adequate lighting for the work area, if natural lighting is inadequate.
3. All workers involved in repairing formwork must wear such PPE as safety helmets, eye protection, safety gloves and foot wear.
4. Safety features of power tools used must be checked before use and extension leads must be protected.
5. All nails and sharp objects should be removed from the formwork material. High tensile nails should be removed with an appropriate tool to prevent nails becoming projectiles when being removed.
6. Progressively remove debris from the work area.
7. Keep the repaired formwork away from access ways.
8. Only persons involved in the repair of formwork should be located in the work area. Any person not involved in the work should be excluded.

Web-based OHS planning system development

In order to reap the best outcome from the proposed knowledge-based OHS planning strategy, the model was translated into a web-based system that

- provides a centralised and commonly accessible means of storing the dynamic OHS knowledge,
- enables the capture and retention of job procedure-based hazards and safe work practices in easily retrievable and exploitable formats and
- allows team members of different sites of a contractor to have access to the central knowledgebase for use and updating.

System modelling

Figure 2.2 depicts the system architecture of the web-based OHS planning system and delineates the various functional components within the system. The system is implemented using a three-tier architecture, which is composed of three logical layers: client side, middle layer and the server side. The user on the client side (tier one) is connected to the knowledgebase on the server side (tier three) through the middle layer (tier two), with access to all the OHS knowledge contents within the knowledge base.

The interface of the system is divided into two major functional sets: (1) knowledge retrieval set and (2) knowledge update set. Users from construction sites can have access to both sets upon access authentication. The knowledge retrieval set allows users from different sites to access the knowledge

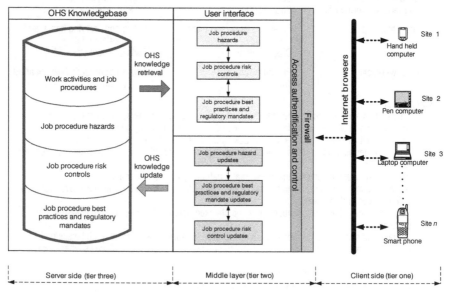

Figure 2.2 Web-based OHS system architecture

about job procedure hazards, safe work practices and regulatory mandates for the job procedure. This knowledge would assist the site personnel in intensive OHS planning on site. The knowledge update set houses templates and forms for updating knowledge on job procedure hazards in the knowledgebase, regulatory mandates and safe work practices in the knowledgebase.

The tree structure in Figure 2.3 illustrates how OHS planning knowledge is retained in the knowledgebase. Possible trades in a construction project are listed on the extreme left. Each trade is then broken down into major tasks and these tasks are further broken down into sub-tasks or job procedures. Subsequently, relevant hazards, risks, root causes and best practices for these sub-tasks are mapped. Knowledge about hazards, risks, root causes and best practices are retained in textual, tabular, pictorial and video formats in the knowledgebase. While texts and tables provide descriptive interpretations of the hazards, risks, causes and best practices, media such as pictures and videos of previous accidents and lessons learned provide interactive explanations that would easily register in the human mind.

Figure 2.4 illustrates the user navigational flow chart for the proposed OHS planning system. The use of the system starts with the Login Page. Once the authentication is verified, the user is allowed access into the Index Page; if not, the user is advised to register. The Index Page directs the user either

Figure 2.3 Knowledge representation tree

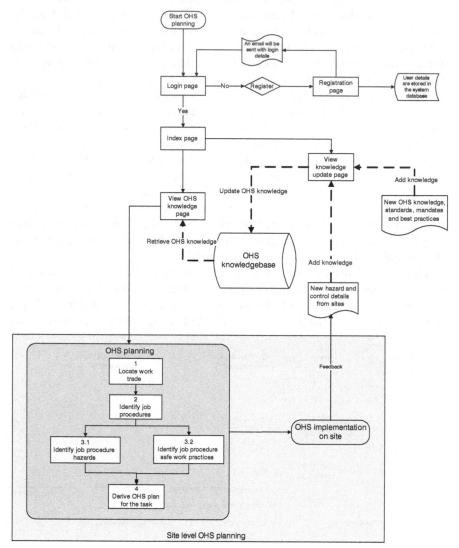

Figure 2.4 Knowledge-based OHS planning flow chart

for: (1) retrieving OHS knowledge for OHS planning or (2) updating OHS knowledge content in the knowledge-base.

Prototype development

A website may be developed by one of the two methods: (1) by coding using web development languages such as HTML, PHP, Cold Fusion, ASP, JSP, etc., or (2) by using a Content Management System (CMS). A CMS is a software program that allows users with little knowledge of programming languages or markup languages to create, edit, maintain and manage dynamic websites using built-in tools and templates. Robertson (2003) suggested that a wide range of benefits can be obtained by using a CMS, including: streamlined authoring process, faster turnaround time for new pages and changes, greater consistency, improved site navigation, increased site flexibility, support for decentralised authoring, increased security, reduced duplication of information, greater capacity for growth and reduced site maintenance costs. Hence, it was decided to use a CMS to make the prototype of the proposed system. There is a plethora of CMS available, both commercially and free general public licences (GPL). Joomla is one of the free CMSs available for researchers and businesses for developing dynamic websites and e-business solutions. The prototype of the proposed OHS planning system was also developed using Joomla. Being a web development tool, Joomla had to be installed on a server and the development task also had to be performed on the server. Server space was leased from a commercial provider and the development was done on it.

How Joomla works

A Joomla website has two parts: a front-end and a back-end. The front-end is that part of the website that is visible to users. The back-end is the administration area where the website is set up, modified and managed. At the back-end, Joomla stores website content in articles, which may contain written information, graphics and/or footages that are added to the webpage. Articles are organised in a hierarchy of sections and categories within the website. An article must belong to a section and a category simultaneously. It is important that initial sections and categories for the website are planned and created before adding content (articles) for individual pages on the website. Joomla also allows there to be uncategorised articles on the website back-end for static pages. Joomla provides a word processor-like visual editing interface for creating and modifying content in articles. Another important element in Joomla is menu manager, which helps create navigational menus for the website. There are ample YouTube videos that explain the creation of sections, categories, articles and menus for Joomla sites. Readers who wish to develop their own websites are encouraged to explore them.

How was the website developed?

The ultimate aim for the proposed system was set to develop a complete e-OHS Management System with four key modules namely OHS Planning, OHS Monitoring, OHS Training and Community of Safety Practice (CoSP) Management. Hence, it was decided to have four major sections for these modules in the initial planning for the website. Subsequently, categories were decided for the OHS Planning section. Construction trades made up the categories, with 14 categories created for the OHS Planning section. Then 11 articles were created for the selected demonstration case of horizontal

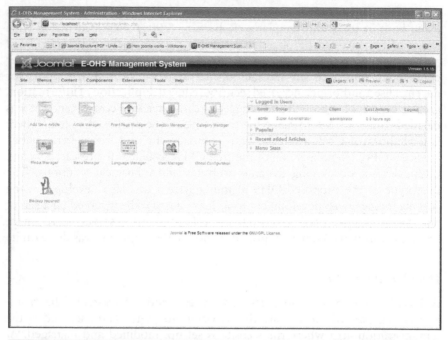

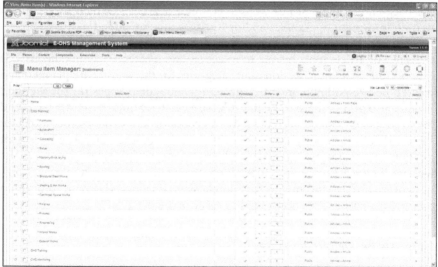

Figure 2.5 Website back-end

formwork in this study, which consists of 11 sub-tasks/job procedures. One article was devoted to describing hazards, risks and best practices for each job procedure and this information is displayed in varying formats such as texts, tables, pictures and video clips. Figure 2.5 depicts the back-end of the proposed web-based system while Figure 2.6 illustrates the front-end of it.

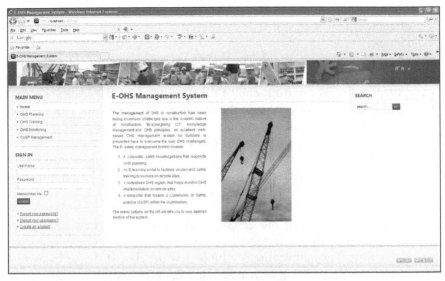

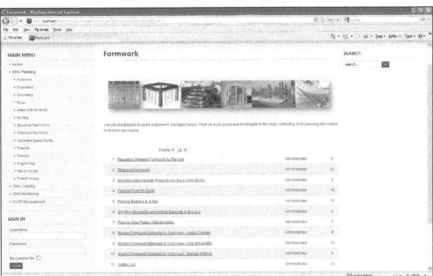

Figure 2.6 Website front-end

How is the website updated continually?

The continual updating of the OHS knowledgebase is facilitated through the access-level configuration in the proposed website. Joomla supports strong access-level configurations for the front-end of the website. There are four front-end user groups available in a standard Joomla website known as Registered, Author, Editor and Publisher.

1 Registered users have the access permission to log in to the website, view all content that is classified as registered access as well as public access content.
2 The Author user group inherits the access permission of the registered user group and, in addition, its members are allowed to create new content items for the front-end of the website.
3 The Editor user group inherits the access permission of the Author user group and, in addition, its members are allowed to edit all published content items for the front-end of the website.
4 The Publisher user group inherits the access permissions of the Editor user group, and in addition, its members are allowed to publish new content items at the front-end of the website.

The continual updating of the OHS-knowledgebase with new knowledge is achieved by providing selected users with author and editor permissions for the website. This allows them submit new articles to the website and/ or add value to the existing contents. Then the administrator, who has the ultimate control of both the back-end and front-end of the website, decides on the suitability of these new addenda for the website. The newly submitted articles may be either left as they are, incorporated into the existing article contents or rejected by the administrator, as the case may be.

Evaluation of web-based OHS planning system prototype

Evaluation exercises for a computer system prototype are carried out to ensure that the system has achieved the outlined research objectives and thus the research hypothesis is validated. In alignment with this broader aim, the evaluation for the proposed web-based system prototype had the following objectives:

1 To demonstrate that the concept and principles that underlie the prototype system could address the challenges facing OHS planning that are outlined in the introduction section of the chapter;
2 To identify aspects of the prototype system and underlying theory that require improvements;
3 To assess the usefulness of the system for its target users;
4 To demonstrate that implementing knowledge management into OHS planning, underpinned by web technologies, could:
   • minimise accidents on site,
   • facilitate learning on-the-job for less-experienced site staff and
   • provide an effective mode for capturing and retaining OHS planning knowledge from different sources for organisational use.
5 To obtain comments and recommendations for further improvements and future developments.

To achieve these objectives, it was decided that potential end-users of the proposed system needed to see a live demonstration of its features and use. They would then be requested to complete a questionnaire that would allow them to express their opinions on various aspects of the system. The evaluation exercise for the proposed system was carried out with 20 representatives from the New South Wales construction industry. The sample size of 20 was considered adequate owing to two reasons: (1) system evaluation exercises for similar web-based systems in previous research suggest that a sample size ranging between 10 and 20 potential end-users would be adequate (Cooke *et al.* 2008; Udeaja *et al.* 2008), and (2) as the evaluation exercise progressed similarities were noted in responses. Figure 2.7 illustrates the profile of the survey participants in terms of their educational background, experience in OHS and type of organisation. Figures 2.7a and 2.7b show the level of OHS competency that the survey participants possess, which suggests that the survey participants had adequate knowledge and experience to evaluate the proposed system prototype. Figure 2.7c illustrates the range of organisations that the survey participants belonged to. This shows that there is an acceptable level of representation from clients, project management consultants, head contractors and subcontractors and therefore it militates against any biased responses.

Questionnaire design

A questionnaire was designed so that the above objectives of the evaluation exercise were achieved successfully. The questionnaire was divided into four sections. Section 1 obtained participants' details. Section 2 evaluated the efficacy of knowledge representation and dissemination approach fostered by the proposed system for OHS planning and accident minimisation. Section 3 analysed the effectiveness of a web-based approach for knowledge-based OHS planning and organisational learning. Section 4 assessed how the proposed system could help overcome challenges facing OHS planning. Additionally, participants were given the opportunity to make further comments under section 2 through to section 4.

Evaluation results

Tables 2.2–2.4 show the average ratings of the system by the survey participants for different criteria. In order to calculate the weighted consensus ratings for each criterion, the following numerical points were allocated for the rating scales:

- strongly agree (SA) = 5
- agree (A) = 4
- mildly agree (MA) = 3
- disagree (DA)=2
- strongly disagree (SDA)=1

a) Educational background

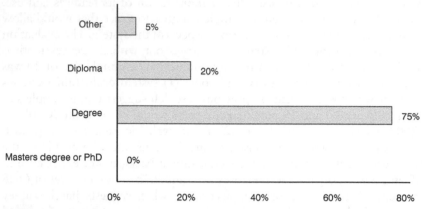

b) Experience in OHS

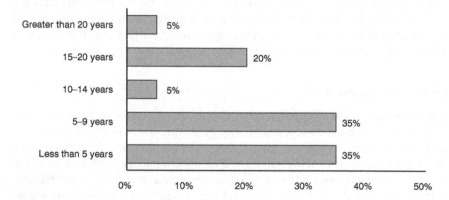

c) Type of organisation

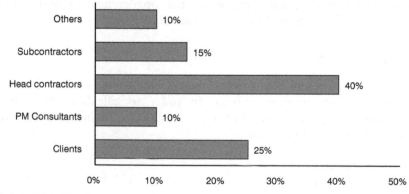

Figure 2.7 Profile of survey participants

Subsequently the following formula was used to compute the weighted consensus rating (WCR) for each criterion:

$$WCR = \begin{aligned}&[(Fraction\ of\ responses\ for\ SA \times 5)\\&+ (Fraction\ of\ responses\ for\ A \times 4)\\&+ (Fraction\ of\ responses\ for\ MA \times 3)\\&+ (Fraction\ of\ responses\ for\ DA \times 2)\\&+ (Fraction\ of\ responses\ for\ SDA \times 1)]\end{aligned}$$

These WCR values explain the strength of consensus by the survey participants for the variable assessed. A detailed analysis of the results for the various sections of the questionnaire is presented below.

Functional features of the web-based system

None of the participants disagreed that the proposed system and the underlying theoretical model were effective methods and could help reduce accidents on site. Further, the weighted consensus ratings for the six variables were greater than or equal to 4.00, which justifies that the conceptual framework of knowledge-based OHS planning and the web-based system are sound approaches.

Benefits of web-based approach for OHS planning

One participant disagreed with variables 2 and 5. The reason mentioned by him is that 'the construction industry still experiences resistance to change from a portion of practitioners and they prefer a paper-based office'. However, 95 per cent of the participants favoured the idea, which is reinforced by the weighted consensus ratings that these variables have obtained. One more participant disagreed that the web-based approach could help improve OHS performance on site. The two reasons he gave were as follows. (1) Despite the system clearly demonstrating the risks and how to deal with them, it is not an enforced system in the industry. If this exercise was enforceable to specific tradesman as well as people overseeing these tradesmen it would increase OHS awareness and practice on site. Hence, the top management should make it compulsory for staff to use the system on site in order to enjoy the benefits of it. (2) Availability and accessibility of the system would be an issue for small-sized contractors and subcontractors, who may not have a computer with internet connection on site. Nonetheless, the rest of the participants believed the proposed approach could help improve OHS performance on site.

Challenges overcome by the web-based OHS planning system

While participants rated many variables in this section favourably, they rated low the suitability of the proposed system for addressing OHS skills

Table 2.2 Effectiveness of the methodology

Functional features of the web-based system	Strongly Agree	Agree	Mildly Agree	Disagree	Strongly Disagree	WCR
1. The demonstrated system is a good method for highlighting the hazards and risks in job procedures to help risk assessments.	35%	45%	20%	0%	0%	4.15
2. The demonstrated system is a good method of explaining the best practices and precautions.	35%	45%	20%	0%	0%	4.15
3. The demonstrated system is an effective and informative tool for OHS planning.	40%	45%	15%	0%	0%	4.25
4. The demonstrated system could help reduce accidents significantly.	35%	45%	20%	0%	0%	4.15
5. Overall, the system could be an effective tool for OHS planning and reducing accidents.	20%	60%	20%	0%	0%	4.00
6. If the system is extended to hold information and videos about job procedures, hazards, risks and best practices for all activities in construction (excavation, concreting, brickwork, etc.), the system could help reduce accidents significantly on site.	35%	50%	10%	0%	0%	4.05

shortage issue among site professionals. The reason they gave was that many site professional still have the attitude that safety can be overlooked, unlike the budget and time objectives of the project. As a consequence, the pressure to finish the work seems to take priority over learning new skills on-the-job.

Further comments and suggestions

Further qualitative comments and suggestions were also provided by the survey participants about the prototype system and these include the following.

- The system is particularly suitable for young tradesmen entering the industry to learn how to do things correctly.
- It is a good tool to assist head contractors to cross-check subcontractors' safety plans. Subcontractors are appointed to carry out trades that do not fall under the specialty of the head contractor. However, the head contractor is required to monitor the safety performance of subcontractors, which poses a challenge as the head contractor may not possess full knowledge about the hazard and risk profiles of those trades. This system may assist the head contractor to address this issue.

Table 2.3 Benefits of web-based approach

Benefits of web-based approach for OHS planning	Strongly Agree	Agree	Mildly Agree	Disagree	Strongly Disagree	WCR
1. The web-based system summarises OHS planning knowledge from different sources (hazard check lists, databases, best practice manuals, codes of practice, etc.), and stores it centrally in easily understandable and exploitable formats.	20%	75%	5%	0%	0%	4.15
2. The web-based approach could make OHS planning knowledge easily available and accessible to all users regardless of their site locations.	45%	40%	10%	5%	0%	4.25
3. The web-based system could save time spent on learning OHS planning knowledge from different sources (hazard checklists, databases, codes of practice, best practice manuals, etc.).	30%	60%	10%	0%	0%	4.20
4. The web-based system could facilitate on-the-job learning of OHS planning knowledge.	30%	50%	20%	0%	0%	4.10
5. It needs no prior training or software skills to use the system but basic knowledge of using web browsers.	35%	50%	10%	5%	0%	4.15
6. Overall, the web-based approach could be an effective way to store and share OHS planning knowledge.	35%	60%	5%	0%	0%	4.30
7. Overall, the web-based approach could be an effective way to facilitate learning on-the-job for site teams.	30%	60%	10%	0%	0%	4.20
8. Overall, the web-based approach could help improve OHS performance on site significantly.	35%	40%	20%	5%	0%	4.05

- The information and video footage could be used as a site tool during inductions of workers and during tool box talks and briefing.
- Users should be able to select specific trades for projects and print out the work method statement from the system, ready to be signed off. It would be good if a pdf report could be created automatically upon selecting the specific trades for a given project, with facilities for adding company and project details in the report.
- The system should be integrated with other web-based project management systems that may be used by builders.
- The framework developed and tested in this research may be used in other industries to improve their OHS performance.
- The system could be used in tertiary institutions to train students on OHS.

Table 2.4 Challenges addressed by the system

Challenges overcome by the web-based system	Strongly Agree	Agree	Mildly Agree	Disagree	Strongly Disagree	WCR
1. The web-based system could help inexperienced site team members to develop better OHS plans.	40%	50%	10%	0%	0%	4.30
2. If the web-based system is populated with more information and videos that cover all activities in a construction project, this could arm inexperienced/less skilled site teams with the necessary skills for effective OHS planning.	40%	50%	10%	0%	0%	4.30
3. The web-based system summarises OHS knowledge from different documents and standards, and makes it easy for site teams to learn and apply, considering their tight schedule and work pressure.	35%	60%	5%	0%	0%	4.30
4. If the web-based system is enhanced with information and videos that summarise OHS knowledge relevant to all activities in a construction project, the system could be a valuable tool for on-the-job learning.	30%	65%	5%	0%	0%	4.25
5. The web-based system could provide a standardised method for capturing, storing and sharing OHS planning knowledge.	25%	55%	20%	0%	0%	4.05
6. If the web-based system is enriched with OHS planning knowledge for all activities in construction, it could be a rich and standardised repository that can be used across fragmented projects of a builder.	20%	65%	15%	0%	0%	4.05
7. The web-based system could help overcome the disparities in OHS planning knowledge among site team members that may arise due to differences in experience and training.	35%	50%	15%	0%	0%	4.20
8. If the web-based system is extended to have information and videos that cover OHS planning knowledge for all activities in construction, the system could help address coverage of knowledge issues with site professionals.	25%	45%	25%	5%	0%	3.90

Conclusions

The development of adequate OHS plans is vital to a construction project of any size and scale to identify hazards as early as possible and implement appropriate control measures. Site management teams face challenges in preparing effective OHS plans owing to factors leading to OHS knowledge shortage among site staff members. The integration of knowledge management, ICT and OHS principles would provide them with the ability to address these challenges. A knowledge-based OHS planning framework supported the development of a web-based OHS planning system. The web-based system was then tested and evaluated in the construction industry by its potential end-user groups. The test results suggest that the proposed system has many advantages for builders.

- It can provide on-demand knowledge that is utilised at the operational site.
- It captures OHS knowledge from different sources, retains and disseminates it to users of it whenever needed, irrespective of their location. It also supports capture of an organisation's knowledge assets created on site.
- It helps users with on-the-job learning of OHS skills.
- It provides with interactive media for safety inductions and tool box talks for operatives on site.
- It could help minimise accidents on site.
- It saves time and money for builders that arise from poor OHS performance.

The proposed system also has advantages for academia whereby it could be used to train students in OHS. The conceptual model behind the proposed system could be used in other industries, with necessary customisations for the industry, to develop similar knowledge-based OHS planning systems, leveraging on web technologies.

3 Affective e-OHS training system for construction

Introduction

Previous studies on causal analysis of accidents in construction suggest that unsafe behaviours of workers have been repetitive causes of accidents in a significant percentage of cases. For example, Haslam *et al.* (2005), upon analysing 100 construction accidents, uncovered that workers' behaviours have been involved in over two-thirds (70 per cent) of accidents. Hinze *et al.* (2005) and Abdelhamid and Everett (2000) also reinforced that a vast majority of accidents on construction sites occur due to one of the following worker-behaviour-related root causes:

- failing to identify and control an unsafe condition that existed before an activity was started or that developed after an activity has started;
- inaccurate perception of risk, with feeling of invulnerability and 'it won't happen to me';
- overlooking safety in the context of heavy workloads and other priorities;
- deciding to proceed with a work activity after the worker identifies an existing unsafe condition; and/or
- deciding to act unsafe regardless of initial conditions of the work environment (taking shortcuts to save effort and time).

The construction industry has been adopting two types of safety training schemes to equip workers with necessary skills and abilities to deal with hazards and act safely on site. These include instruction-based (standard) safety training and behaviour-based safety training. While the former is a classroom-based approach, the latter is a worksite-based approach. The standard safety training aims at safety capacity and knowledge building for workers. All new workers receive the same basic safety training which is predominantly classroom-based and consists of reading materials and lectures supported by audio-visual aids such as video tapes and power point presentations. Behavioural safety is an approach to safety training that specifically focuses on the safe and unsafe behaviours of workers in the workplace. Its aim is to identify, influence, modify and change unsafe behaviours of workers in the workplace. It also reinforces the safe behaviours of workers that have already been established. At the core of this training programme are behavioural observations, behavioural feedback and behavioural interventions. The behavioural observations involve observations being conducted on an individual or a small team of people who are performing a task. The purpose is to determine what is done safely and what unsafe behaviour is. The behavioural feedback involves the observer providing the observee with balanced feedback on the safe and unsafe behaviours in a positive and constructive manner on completion of a behavioural observation. The behavioural interventions involve rewarding safe behaviours through incentives and providing safety training on specific issues identified in a behavioural observation.

Nonetheless, the occurrences of accidents owing to unsafe behaviours of workers still seem persistent. This could be due to the fact that the current training schemes operate at a higher level and desire to alter workers' behaviours through the acquisition of skills and awareness in safety. In order to achieve better safety outcomes, however, it is equally critical to go a level below to analyse and treat a primitive factor, the attitude of workers towards safety. Attitude is an internal state of mind that determines the behaviour or performance of individuals and there is evidence that attitude can be changed through a variety of media (Kraiger *et al.* 1993). Based on this definition, it is clear that unsafe behaviours of workers on construction sites can be changed by changing their attitudes towards hazards and safety. The deployment of a mentally influential medium is crucial to achieve the desired attitudinal change in workers. Gano Phillips (2010) maintained that 'affective' education and training influence learners' attitudes, values and motivation. These produce successful learning outcomes, including: (1) showing awareness and selective attention to issues, (2) accepting responsibility for one's behaviour and actions, (3) active participation and willing responsiveness, (4) accepting professional ethical standards and (5) demonstrating ethical behaviours in accordance with guidelines in activities. Findings in neuroscience, psychology and cognitive science also recognised that affect plays a crucial role in guiding rational behaviour, memory retrieval, motivation, attention, decision-making and creativity (Picard *et al.* 2004). Affective OHS training for workers should help address the challenges posed by unsafe behaviour-related root causes of accidents and thereby minimise accidents considerably on site.

To this end, this chapter discusses the development of a web-based interactive system for affective OHS training for workers that could influence workers' attitude towards safety and thereby safe behaviours. The discussion is laid out logically in the following order. First, the theory of affective learning is explored followed by tools for affective learning. Then an affective OHS training model for construction is proposed, followed by discussions on the development and validation of a web-based interactive tool that implements the proposed training model in a practical sense.

Affective learning theory

Oxford English Dictionary defines affection as 'emotion or desire as influencing behaviour'. Affective learning involves the melding of thinking and feeling in how people learn. Importance is placed on social learning environments for knowledge construction and application wherein deeper awareness and understanding of the role played by mental dispositions in how a person views, engages and values learning can result in better understanding and use of knowledge and skills. Learning outcomes are focused on enculturation of norms, values, skilful practices and dispositions for lifelong learning (Stricker 2009). Affective learning theory has gained more attention due to the emphasis on a humanistic approach in education. The humanistic approach emphasises the individuals' feelings and emotions in the teaching and learning process, in other words, they are human beings who have a natural potential for learning (Wang 2005). Cognition and emotion can be seen as partners in the mind, thus they cannot be separated from each other to construct a firm foundation in the learning process (Meyers and Cohen 1990; Goldfayl 1995; Jones and Issroff 2005; Wang 2005). Learning occurs when learners are involved emotionally, therefore, integration of the affective domain into teaching and learning methods is critical. Tooman (2010) claimed that it seems vital to stimulate the affective factors for adult education because it helps adults draw meaning out of life experiences.

'Affect' refers to a full range of learners' emotional responses, both positive and negative (Goldfayl 1995). Various researchers have suggested different lists of basic affects that influence learning. Ekman and Friesen (1978) focused on six basic affects/emotions: fear, anger, happiness, sadness, disgust and surprise. Plutchik (1980) identified eight affects/emotions, including fear, anger, sorrow, joy, disgust, acceptance, anticipation and surprise. Tomkins (1981) proposed nine critical affects: interest, enjoyment, surprise, fear, anger, distress, shame, contempt and disgust (Tomkins 1981; Goldfayl 1995). The four most common affects/emotions appearing on the many theorists' lists are fear, anger, sadness and joy (Kort *et al.* 2001). Researchers have found that, while positive emotions such as interest, enjoyment and surprise have a positive effect on learning and social behaviour, negative emotions such as anger and sadness have a negative impact on learning and motivation. However, negative affect such as anxiety or fear can help individuals to focus, leading to better concentration and reducing distractions (Moridis and Economides 2008).

In the context of OHS training, it could be derived that the utilisation of training resources which are deliberately designed to simultaneously stimulate anxiety or fear of death, incapacitation and/or injuries in workers, while delivering the learning contents, could result in better concentration and cognition by workers. This would eventually result in permanent safety consciousness in the minds of workers that would influence their behaviours on site. Figure 3.1 presents this notion graphically for easy comprehension.

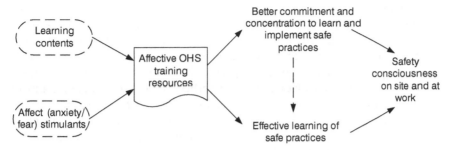

Figure 3.1 Affective learning of safety

Tools for affective learning

Learning contents and delivery modes have significant influences on affective learning. Herreid (1997) examined the case studies method of teaching and concluded that stories, as a means of teaching cases, touch human beings' fundamental nature, which can produce excitement, sadness and other emotions. Stories, being the oldest and most natural and powerful form for storing and describing experiential knowledge, are 'means by which human beings give meaning to their experience of temporality and personal action' (Jonassen and Hernandez-Serrano 2002). Stories can be used to support learning, providing rich examples. Stories are also embedded with instructions which advise us in our complex life. Case studies are stories with an educational message and have been used as parables and cautionary tales for centuries (National Center for Case Study Teaching in Science 2010). Thus, incident case libraries, collections of incident cases and experiences can be valuable tools for OHS training. Charles *et al.* (2007) advocated in their guide to best practice for safer construction that stories of previous incidents should be utilised for OHS training with appropriate reinforcements to enhance workers' engagement in learning and application. Factors that could be integrated as reinforcement in the cases may include human behavioural failures that led to incidents and human suffering (death, incapacity, injuries, etc.) as a consequence of it, and safe practices that could have prevented the human misery.

Herreid (1997) suggested some basic rules for creating good case libraries for affective OHS training. A good case (1) tells an informative story, (2) focuses on an interest-arousing and important issue, (3) should instil fear or anxiety of death and/or incapacity in learners' hearts and minds, (4) is relevant to the learner and applicable in real life, (5) must have pedagogical utility, (6) is thought-provoking and forces the learners to take a position in the case, (7) is short, introducing the facts with enough time but not making the learners bored and (8) includes quotations that add life to cases and give realism. Furthermore, Life Lines Online (2010) described various formats of cases for different learning objectives, including:

1　Extensive, detailed cases – focus on a particular decision, the people who made it, the people affected by it and the impact of that decision on all parties. Frequently used in business courses with 100 pages or more.

2　Descriptive, narrative cases – designed to be used over the course of two or more class meetings, with up to 5 pages. The case is disclosed to the learners one page at a time with discussion, hypothesis generation and development of learning goals and study questions for each part of the case.

3　Mini cases – designed to be used in a single class meeting and tightly focused. These cases are useful for helping learners apply concepts and introducing practical applications.

4　Bullet cases – two or three sentences with a single teaching point. Learners discuss them in small groups.

5　Directed case study – short cases are followed immediately with highly directed questions.

6　Fixed choice options – a variation on 'bullet cases', a mini case with four to five plausible solutions which is useful for policy, ethics and design decisions. Learners must choose and defend one solution.

The development of a case library with mini cases of incidents, with consideration of Herreid's eight rules above, could help achieve the desired outcomes in affective OHS training.

Different types of learning environments and contents could result in different affective learning. The internet and multimedia can provide complex and real learning environments, particularly for case studies which usually have rich contents. Multimedia means a combination of text, audio, images, animation, video and interactive content forms. Oller and Giardetti (1999) stated that, to capture attention, nothing works better than pictures, and moving pictures are better than still pictures. Video, with graphic and text-based support, provides a more real learning environment for both learners and teachers. Compared to text-based cases, real video allows for different interpretations, reflections and analysis (Lin 2001). Racicot and Wogalter (1995) found that behavioural changes induced by video warning signs are robust over time. A real-world case study can be communicated as a video story, illustrating what has happened in the past and what the outcome was for that event and adding authentic context and emotional appeal (Snelson and Elison-Bowers 2009). The internet makes new media, like online videos, possible for delivering to learners, and also provides 'three significant capabilities' for the delivery of case studies (Kovalchick *et al.* 1999): (1) the ability to stimulate real-world complexities, (2) the ability to use multimedia in case presentations and (3) the ability to use hyperlink/hypertext navigation features

Affective OHS training model for construction

In the light of paradigms and tools explored above, an affective OHS training model is proposed as a means for cultivating a positive attitude towards safety in construction workers. Figure 3.2 illustrates a conceptual model of the affective OHS training scheme that is advocated in this chapter. As can be seen in the diagram, the model is composed of four key processes.

1   Undertaking effective incident investigations and codifying informative incident cases so as to be exploitable for OHS training of workers;
2   Reconstructing affective incident case videos with reinforcements of affects and safe practice instruction;
3   Establishing an affective OHS training courseware that consists of a training case base and content retrieval engine; and
4   Accessing the courseware through the internet and administering context-specific affective OHS training on site to workers.

The notion of affective OHS training that is promoted through this model is heavily reliant on leveraging on past incidents for altering workers' unsafe behaviours on site. It is therefore critical to carry out effective incident investigations and capture information on: (1) the location and work activity where the incident occurred, (2) the nature of incident, (3) the causes of incident, with emphasis on workers' unsafe behaviour, (4) the details of injuries/death/incapacity suffered and (5) suggestions for preventing recurrences. The incident case with the above information may be structured and stored in an incident case base that can be referred to for reconstructing videos of the incidents as explained in the next section.

The second process in the proposed model involves the reconstruction of affective accident case videos. Incidents are hardly video-recorded at the time of occurrence and therefore it is almost impossible to have real video footages of incidents for training purposes. However, incidents may be reconstructed with the aid of multimedia technology. This creates ample opportunities for embedding OHS pedagogical utilities and affects for educating workers. When reconstructing case videos emphasis should be placed on three aspects:

1   Embedding and highlighting unsafe behaviours of workers that led to the incident and the nature of incident;
2   Highlighting the injury severity or suffering to the victims of the incident in such a way that it instils anxiety and fear in the minds and hearts of observers; and
3   Embedding safe practices that should be followed to avoid similar incidents on workers' own worksites.

When cases are reconstructed in this fashion, they could help create an attentive learning environment for workers. This could, in turn, reap better OHS training outcomes. The information captured during the incident investigation phase can be utilised to embed the first two aspects in case videos. In order to embed safe construction practices, builders may use their own knowledgebase or other published information, as the case may be. However, the proposed system utilises both the training package provided by International Labour Organisation (ILO) for worker training and Australian Codes of Practice for construction work. The reason for using the ILO training package is that ILO has developed this training package after in-depth research and analyses of accidents across many countries. Thus it tends to address issues in specific details. Moreover, the package is relevant to a global audience, is designed to be applicable in differing political, cultural and legal environments and it is internationally recognised (ILO 2010). The Australian Codes of Practice were utilised together with the ILO training package to ensure the contents of the training videos cover the local context in the local industry language. A detailed account of the knowledge content in the ILO knowledgebase and the Australian Codes of Practice and how it is utilised in the construction of affective incident case videos are presented in the next section.

The next process in the proposed model is the establishment of an affective OHS training courseware. With the reconstructed incident case videos, an e-OHS training portal can be set up that can be accessed remotely, around the clock. The portal will essentially have two components, namely an affective OHS training case base and content retrieval engine. The training case base organises affective incident case videos in a logical structure so that it will help context/activity-specific OHS training for workers. This chapter adopts the structure provided in the ILO knowledgebase to organise the case videos in the system which is discussed in the upcoming sections. The content retrieval engine of the courseware would provide users with a fast case-retrieval, based on set search criteria through a graphical user interface.

Once the OHS training courseware has been established in the organisation, it can be used online by work supervisors to support OHS training, such as providing context-specific hazard alerts to workers prior to commencement of an activity and tool box talks. It could even be used to underpin the standard training. The courseware may be accessed using portable digital devices and/or smartphones. However, considerations should be given to bandwidth and speed of video play when reconstructing the case videos for the courseware.

Elicitation of OHS knowledge for affective training

ILO recognised that construction deaths worldwide run at over 100,000 per year – that is, one person every five minutes. Decent safety training for all those working in the construction industry is desperately needed to minimise this. However, ILO noted a shortage of comprehensive construction safety training materials in the public domain. In order to fill this gap, it developed a training package, called *ILO Construction Occupational Safety and Health* (OS&H), to provide trainers/supervisors with materials required to plan,

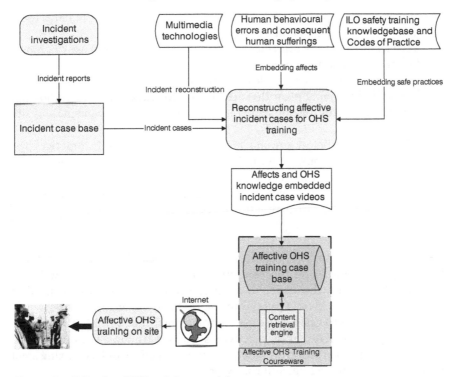

Figure 3.2 Affective OHS training model

create and deliver a construction safety and health course, built to the needs of workers (ILO 2010). Various codes of practice for construction that are published by Safe Work Australia and WorkCover NSW also supplement the ILO training package. This section deals with the elicitation of knowledge on safe construction practices from the ILO training package and the other sources and representing it in an abstract format. This would produce a set of structured pedagogical contents that can be used for reinforcing affective training videos.

The ILO training package recognises that workers should be trained in OHS precautions for seven activities during common construction operations, and each activity has subsections so that most construction operations are covered in the training package.

- Welfare and project site – expounds guidance on safe welfare, site layout, housekeeping, warning signage and display, site security, storage and temporary services.
- Personal protective clothing and equipment – presents safe and effective use of personal protective clothing and equipment to deal with such hazards during construction operations as impacts, heat, chemicals, dust, optical radiation, noise and heights.

- Power and handheld tools – explains safe techniques to use hand tools and other power tools that are used on site.
- Vertical movement – describes safety considerations in the use of cranes, hoists and concrete pumps.
- Horizontal movement – deals with safety precautions when using horizontally moving plant that are utilised for excavation, earthmoving, compacting, road-making, concrete production and movement, and site transport.
- Working at or below ground level – elaborates on safe practices for common excavations, shafts and headings, works on contaminated sites, demolition and works in confined spaces.
- Working at heights – covers safe practices for ladders, scaffolding, structural frame construction (formwork, reinforcement and concreting) and roof work.

Nevertheless, due to time constraints, this research demonstrates the knowledge elicitation and representation only in the context of 'working at heights'. Readers who desire to learn the complete contents of the ILO training package are advised to refer to the ILO knowledgebase (http://www. ilo.org/public/english/dialogue/sector/sectors/constr/index.htm). Working at heights largely involves six key activities on construction sites, including ladder use, scaffold erection and use, formwork erection and use, fixing steel reinforcements, concreting, and roof work.

Ladder safety

Fifteen essential ladder safety instructions have been drawn from ILO (2010) and Safe Work Australia (2010) and are presented below.

1 Always inspect your ladder before use – check the ladder is in good condition and strong enough for the job.
2 Check if the ladder is tall enough for the job.
3 Make sure to place the foot of the ladder on a firm and level base.
4 Place the ladder at a safe angle – for every 4–up, place the ladder base 1–out from the wall.
5 Secure the ladder at the bottom by tying it to stakes in the ground or by using sandbags.
6 Rest the ladder against a solid and strong surface or use a ladder stay.
7 The ladder should extend at least 1m above the landing place.
8 Secure the ladder at the top.
9 Make sure there are no overhead powerlines with which the ladder might make contact.
10 Check that your footwear and ladder rungs are free of grease and mud before climbing the ladder.
11 Do not carry tools or materials in your hand while climbing the ladder.

12 Keep three points of contact with the ladder at all times – two limbs and one hand or one limb and two hands.
13 Do not over-balance or over-reach.
14 On extension ladders, make sure two sections overlap adequately according to manufacturer's instructions and engage all locks properly in the overlap.
15 When using step ladders, fully open the step ladder and lock spreaders in place. If possible brace yourself with your free hand.

Scaffold safety

Twenty-three scaffold safety instructions have been drawn from ILO (2010) and WorkCover (2001a) and are presented below.

1 Make sure you and others that are involved in erecting scaffolds are ticketed (certified) for the task.
2 Check if the ground or foundation of the scaffold is level and stable before erecting and/or climbing a scaffold for work.
3 Make sure the materials used for erecting scaffolds are of adequate type, strength and size for the job (light duty or heavy duty).
4 Do not use damaged parts and components to erect scaffolds.
5 Never use loose bricks, drain pipes or other unsuitable materials to erect or support any parts of scaffolds.
6 Brace the scaffold adequately both diagonally and laterally.
7 Where applicable, rigidly connect the scaffold to the structure at suitable vertical and horizontal distances.
8 Do not exceed the recommended tower height for the scaffold.
9 Set the planks of the working platform at the same surface height and close any gaps between planks.
10 Provide guardrails and toe boards around the scaffold working platform and access stairways.
11 Provide overhead screens of adequate strength and dimension for scaffolds to prevent danger from falling objects.
12 Do not use cross braces of the scaffold as a means of access. Use a secured ladder, ramp, stair tower or prefabricated scaffold access.
13 Never erect, work on or move scaffolds too close to live powerlines or unprotected light bulbs.
14 Do not use scaffolds too close to plant operating areas. Define boundaries of scaffold use clearly.
15 Do not lift heavy and large members or planks alone; practise team lifting.
16 Never use ladders on any working platform to extend the height of the scaffold.
17 Do not move a scaffold tower with materials or workers on it.
18 Do not throw materials or tools from scaffold working platforms.

19 Do not use scaffolds on sloping or uneven grounds.
20 Avoid striking against the scaffold when lifting and transferring heavy loads on to the scaffold.
21 Do not overload the scaffold platform with materials and/or workers. Evenly distribute the load on the scaffold platform.
22 Keep the scaffold platform free of debris, oil and/or grease. Do not work on a slippery scaffold platform.
23 Always use appropriate personal protective equipment during the erection of scaffolds and working on them.

Formwork safety

Twenty-one formwork erection safety instructions were identified form ILO (2010) and WorkCover (2001b) and are summarised below.

1 Never start formwork erection until the open edges of the floor/work area are protected by scaffolds/guardrails or you are provided with a safety harness.
2 Do not work on upper floors or plydeck in extreme weather conditions such as rainfall, storm, lightning or gusts of wind.
3 Do not stack formwork materials across access ways and manoeuvre routes.
4 Remove all pre-existing nails and other sharp objects from formwork frames, joists, bearers, planks and plywood before using/handling them.
5 Use only non-faulty planks of adequate size and strength for working platforms and support them securely at both ends.
6 Do not climb up formwork frames/braces to reach working platforms/ plydecks. Use a safe means of access like a ladder.
7 Do not lean outward from working platforms to fix frames, braces, bearers or deck.
8 Never walk on unsecured joists.
9 Never work close to an unprotected penetration/hole in the floor or plydeck. Guard/cover all penetrations/openings (hatchway, chute and/ or trapdoor openings) before starting work.
10 Keep the working platforms and plydecks free of dusts, oils, off-cuts, grease, nails, etc. at all times.
11 Only lift formwork frames, joists, bearers, planks and plywood of manageable weight and size at heights. Avoid lifting them at heights when it is too windy.
12 When passing up bearers, joists and plywood up to co-workers, only handle members/planks/plywood of manageable size and weight.
13 Practise job rotation and team lifting and use lifting aids.
14 Do not throw tools up to co-workers. Use a tools bucket to pass tools up.

15 Before using power tools such as nail guns, saws, etc., for formwork erection, make sure extension leads, plugs and power tool parts are free of fault and guarded.
16 Do not connect several extension leads together.
17 Make sure earth leakage switch is installed properly to the mains supply.
18 When using power tools (nail guns and saws), keep your hands, fingers and legs away from the power tool operating area.
19 Barricade formwork erection zone, install warning signs and deploy a spotter to prevent trespassers.
20 Always wear PPE provided, such as safety boots, gloves, sun protection clothing, ear protection, safety harness, eye, ear and respiratory protections at all times without fail.
21 Protect elevator shafts, chute walls and window wall openings with a standard guardrail on the upper floor where a formwork frame is erected.

Reinforcement safety

Twenty-one safety instructions for fixing reinforcement were elicited from WorkCover (2001c) and are presented below.

1 Never work on unprotected plydeck for fixing reinforcement unless you are provided with a safety harness.
2 Do not work close to penetrations and deep beams on plydeck. Guard/cover them before the work starts.
3 Do not clutter the plydeck with steel bars. Stack them uniformly without obstructing access and worker manoeuvre routes.
4 Keep the plydeck free of dust, debris, steel off-cuts, sharp objects, oil, grease, etc. at all times.
5 Do not work in poorly lit areas.
6 Use enough manpower for carrying around and fixing large steel bars in position.
7 Do not walk on steel bars or steel cages.
8 Do not walk backward when placing steel.
9 Always alert co-workers when carrying around and/or unloading steel bars.
10 Stay away from the steel load landing area when cranes are used for lifting steel bars onto plydecks.
11 Always wear flexible leather gloves, long pants, boots and eye protection when handling steel bars.
12 Cover, remove or bend down all sharp edges of steel bars and tie wire.
13 Have sufficient rest breaks or job rotations when performing repetitive bending, twisting and cutting with nips.

14 Always wear protective clothing – sunscreen 15+, shirt and flap-on hard hat.
15 Continually rehydrate yourself with cold drinking water in hot weather.
16 When cutting steel bars using angle grinder, drop saw or Oxyacetylene torch, clear all combustible materials from the work area prior to cutting and keep a filled-up fire extinguisher adjacent to the work area.
17 Before using angle grinder, drop saw or Oxyacetylene torch for cutting steel bars, make sure extension leads, plugs and power tool parts are free of faults and guarded.
18 Do not connect several extension leads together.
19 Make sure earth leakage switch is installed properly to the mains supply.
20 When using power tools keep your hands, fingers and legs away from the power tool operating area.
21 Do not work on upper floors or plydeck in extreme weather conditions such as rainfall, storm, lightning or gusts of wind.

Concreting safety

Twelve concreting safety instructions were derived from ILO (2010) and WorkCover (2001d) and are listed below.

1 Never work on an unprotected floor for concreting unless you are provided with a safety harness.
2 Do not work close to penetrations in a concreting work face. Guard/cover them before concreting starts.
3 Do not perform concreting in a dark or poorly ventilated work face.
4 Stay away from delivery hose at the first moment of delivery at start/restart of the pour.
5 Make sure parts of delivery hose, vibrator and float are free of fault before using them for concreting.
6 Use sufficient manpower for lifting and operating concreting machinery such as delivery hose, vibrator or float.
7 Do not walk backwards while pouring, shovelling, vibrating, trowelling and screeding concrete.
8 Do not walk behind operators of delivery hose, vibrators or float.
9 Do not step in gaps between reinforcement bars. Walk on meetings of bars instead.
10 Have enough rest breaks or job rotations during shovelling, trowelling, screeding or vibrating concrete.
11 Never smoke near the concreting area.
12 Always wear safety clothing, ear, eye and respiratory protections provided.

Roof work safety

Twenty-five roof work safety instructions were identified from ILO (2010) and WorkCover (2001e), and are presented below.

1 Do not start roof work until the edges of the roof area are protected by guardrail or you are provided with a safety harness.
2 Do not work on a roof in extreme weather conditions such as rainfall, storm, lightning or gusts of wind.
3 Use safe means of access (ladders, scaffold, stairways or mobile hydraulic platforms) to the roof area.
4 Never stand or sit on skylights or similar brittle roofing materials. Use a crawling board/roof ladder to span across skylights.
5 Do not walk in the middle of battens. If necessary walk on the rafter/batten joints only.
6 Do not lift large or heavy materials alone on the roofing area. Use team lifting.
7 Keep the roofing area free of slippery substances such as oil, grease, dew, water, debris, etc.
8 Wear appropriate footwear, gloves and sun protection at all times.
9 Before using power tools such as nail guns, saws, etc. on roofs, make sure extension leads, plugs and power tool parts are free of fault and guarded.
10 Do not connect several extension leads together.
11 Make sure earth leakage switch is installed properly to the mains supply.
12 When using power tools (nail guns and saws), keep your hands, fingers and legs away from the power tool operating area.
13 Wear the ear, hand, eye and respiratory protection provided without fail.
14 When lifting materials using a crane, make sure the dogman and crane operator are qualified for the operation.
15 Use proper sling/chain for the materials and secure the sling correctly.
16 Never overload the sling.
17 Be watchful of the sling swing area so that it does not strike you.
18 Maintain good communication and coordinate the lifting with the dogman and crane operator.
19 Disconnect all power connections to the roofing area except the temporary power supply for tools.
20 Make sure insulations are installed on overhead powerlines above the roof before starting the work.
21 Do not touch/strike overhead power lines with batten or other roofing materials.
22 If possible, cut roofing members and sheets on the ground.
23 Do not fail to communicate with people on the ground before dropping materials or objects.
24 Remove all the off cuts immediately from the roof area.
25 Barricade drop zone, install warning signs and deploy a spotter to prevent trespassers.

1. Worker's unsafe behaviours that caused the first incident are highlighted. 	2. The mishap of falling off the ladder is shown. 
3. The preventable fatal consequence of the unsafe behaviour of the worker is brought to light. 	4. Causes are re-iterated for relating them to precautions that could have saved this worker. What caused this unnecessary death? 1 The ladder was too short for the job 2 The ladder wasn't secured at the bottom and top 3 The angle of the ladder was insufficient 4 The extension of ladder above the place was too short 
5. Subsequently, nine ladder safety precautions are explained. 	

a)   Images 1-3 © WorkSafeBC and used with permission; images 4–5 Vocam and used with permission.

Figure 3.3a Affective OHS training videos for ladder safety

Reconstruction of affective incident case videos for OHS training

The reconstruction of affective incident case videos involved three stages:

1 Reconstructing incident case videos vividly highlighting the incident, severity of injury to workers and causes of incidents. These videos were reconstructed in such a way that the injury parts instil fear/anxiety in viewers' hearts and minds.

2 Developing video explanations of safe practices and video warnings based on the OHS training knowledge elicited above. These visual media are able to transmit the message to workers more effectively than verbal

1. Workers' unsafe behaviours and misjudgements of hazards that caused the second incident are explained. 	2. Electrocution of worker is shown. 
3. Fall of a worker to death from the ladder due to electric shock is elucidated. 	4. Causes of the death are textually reinforced. What caused the deadly electrocution and fall of the worker? 1 The presence of the overhead power line wasn't considered 2 The gutter was too long 3 The worker was carrying the gutter on the ladder 

5 Then six ladder safety precautions are explained.

b)   Images 1-3 and 5 © WorkSafeBC and used with permission; image 4 Vocam and used with permission.

Figure 3.3b Affective OHS training videos for ladder safety

explanations and capture workers' attention easily. Moreover, they can address language barrier problems during OHS training.

3 Embedding relevant video explanations of safe practices and video warnings after highlighting the cause of incidents, as possible precautions that could have eliminated the causes and prevented the incidents. This helps workers relate and contextualise the knowledge gained and apply it to similar contexts on site in future.

Figure 3.3a and b depict affective OHS training videos for ladder safety, as an exemplary demonstration. These videos were built based on two ladder incident stories and embed the fifteen ladder safety instructions for OHS training purposes. The first incident story reports on the fall of a worker to death from an unsecured ladder, which was also too short for the job. Nine ladder safety instructions are linked to this incident for training purposes. Likewise, the second story discusses the fall of a worker to death from a ladder due to combined causes of electrocution and carrying materials on ladders. The remaining six ladder safety instructions are linked to this incident for OHS training of workers. Figures 3.3a and 3.3b show some screenshots from the affective training videos.

Cells 1 to 3 in Figure 3.3a narrate the first incident, that is the fall of a worker to death from an unsecured ladder. Cell 1 describes how the incident started and establishes the causes that led to the incident. Cell 2 shows how the incident actually occurred while cell 3 vividly highlights the death suffered by the worker due to his ignorant unsafe behaviour. Cell 4 reinforces the causes of accidents textually. Subsequently, nine out of fifteen ladder safety instructions are explained as a means to eliminate possible recurrences. Cell 5 shows two of the safety instructions. Likewise, cells 1 to 3 in Figure 3.3b relate the second incident that is the fall of a worker to death from a ladder due to electrocution. Cell 1 highlights the unsafe behaviours of workers and establishes the causes of the incident, while cells 2 and 3 show how the incident occurred and what the human suffering was. Akin to the previous video, cell 4 here too reinforces the causes of incident textually. Finally, the remaining six ladder safety instructions are elaborated as preventive measures to avoid possible recurrences on site. Cell 5 shows two of the safety instructions.

Similar procedures were adopted to develop affective OHS training videos for other activities such as scaffold erection and use, formwork erection and use, fixing steel reinforcements, concreting and roof work. Because these videos are meant for an online courseware, considerations were given to the type and display quality of video for online streaming during the reconstruction. The play duration of the videos was limited to a maximum of four minutes in order to retain the attention of viewers.

Web-based training courseware development

The affective OHS training courseware was developed using Joomla, which is a content management system. Joomla was chosen due to its many advantageous features, including:

- it is free to use and therefore a very viable option for small and medium sized builders,
- less programming expertise is required to develop a system on Joomla,
- it provides easy to use layout settings, page templates, plug-ins and interfaces,
- it is easy to integrate videos,

Table 3.1 Website layout map

Navigation mapping	Contents included
• Home	General introduction to unsafe acts and their consequences.
• Contractor's responsibility	This page contains video resources, titled Contractor Fined and Duty of Care. These videos explain the OHS responsibilities of a main contractor and subcontractors and the consequences of negligence.
• Workers' rights and responsibilities	This page has three videos covering: (1) consequences of bullying and how to deal with it, (2) consequence of not reporting hazards and (3) site housekeeping tips.

• Works at heights safety
  • Ladder safety
  • Scaffold safety
  • Roof work safety
  • Formwork safety
  • Reinforcement safety
  • Concreting safety
	The main menu Works at Heights Safety shows an affective video in that the story of a wheelchair-bound ex-construction workers is narrated.
	The submenu Ladder Safety links to a page that contains ladder-related accidents and ladder safety precautions. Likewise, the submenu Scaffold Safety is connected to a page that contains scaffold-related affective accident videos and then scaffold safety precautions. Other submenus have been designed in the same way.

• Hand and power tool safety • Vertical lifting plant safety • Mobile plant safety • Excavation safety	No contents were included for these menus as this was beyond the scope of the discussion this chapter. However, these were included as menus as a part of the complete conceptual design for the affective OHS training courseware.

- sites can be built collaboratively with others, and
- they are fast and easy to build.

Being a web development tool, Joomla had to be installed on a server and the development task also had to be performed on it. Server space was leased from a commercial provider and the development was done on it, which was a cheaper and hassle-free option for the researcher.

A site layout map was first designed that describes the contents and navigation links that form the menus on the website. Table 3.1 explains the layout and the interactive content that goes under each menu and submenu in the system interface. The menu names have been derived based on the training category proposed by the International Labour Organisation.

Figure 3.4 depicts a screenshot of the proposed web-based OHS training courseware. The menus on the sidebar reflect subjects for OHS training. Works at heights is one of them and the demonstration case for this research. Affective training videos can be viewed by navigating through relevant menus and submenus in the system. For example, if a supervisor wants to show affective training videos of roof work safety to workers at site inductions, the videos can be located by following the path of 'Works at Heights>Roof Work Safety'.

Figure 3.4 Web-based OHS training courseware. Images are ©WorkSafeBC and used with permission

Effectiveness of the courseware

An evaluation exercise was administered to assess the effectiveness of the proposed affective OHS training scheme along with its web-based platform for achieving its intended goals. The evaluation intended to assess if the new training model would be helpful to:

1 Influence positive attitude changes in workers;
2 Create an attentive learning environment;
3 Present OHS training contents in fool-proof format; and
4 Minimise unsafe acts of workers and thereby accidents on site.

Potential end-users of the proposed system were approached for evaluation. A potential end-user was defined as construction site professionals who are involved in workers' safety training. Tan (2007) claimed that personal interviews are most suited if probing questions are involved or visual demonstrations are required. In the evaluation exercise for the proposed system probing questions needed to be asked subsequent to visual demonstrations of the proposed web-based system. Hence, this approach was considered more appropriate to obtain site professionals' opinions on the proposed system. The evaluation exercise aimed to understand the views of site professionals on five key issues regarding the proposed training system and therefore responses were sought for the following five interview questions.

Table 3.2 Profiles of construction professionals surveyed

Position	Experience in construction OHS	Type of organisation
OHS coordinator	> 10 years	Main contractor
National OHS manager	>10 years	Main contractor
Group OHS manager	5–10 years	Main contractor
OHS manager	5–10 years	Subcontractor
Site manager	>10years	Subcontractor
OHS risk management officer	5–10 years	OHS training institution
Construction manager	5–10 years	Subcontractor
Construction supervisor	> 10 years	Subcontractor
Construction supervisor	5–10 years	Subcontractor

- Could the proposed web-based system help to improve workers' attitude towards safety?
- How might this web-based training system be utilised?
- What benefits might be gained by using the new training scheme?
- What measures could be taken to harness the usage of the training scheme in construction?
- What obstacles might be faced for its successful implementation?

Participants

The interview survey involved nine construction professionals who deal with site safety and construction workers on a daily basis and Table 3.2 shows their details. The survey participants have adequate exposure to and experience in OHS training for workers and understand the inherent issues and challenges. This is also reinforced by their profiles in Table 3.2. In the evaluation interviews, the proposed system was demonstrated to potential end users and then responses were sought from them to the above questions. The responses given by the participants were constantly positive and similar, endorsing the effectiveness of the system for the intended purpose. Therefore, the evaluation exercise was concluded with nine professionals.

Findings

The responses provided by participants were collated and qualitatively analysed. The summary of findings is presented here.

Improving workers attitude towards safety

The participants unanimously consented that the proposed training scheme would be a better approach to changing workers' negative attitude towards safety. Some of the participants are quoted from here:

The training videos will affect the attitude of workers and constantly refreshed and reinforced in their minds if they are used on an ongoing basis. Most training programmes fail because of workers who believe 'that won't happen to me'. Continual utilisation of this system for training could help soften this type of mindset too.

The videos in the proposed system are hard hitting and will help to move workers. Explaining safety precautions and other associated information to workers after drawing their attention via these hard-hitting videos is an effective approach.

The combination of visual media and balanced course contents can draw the attention of workers and are better stimulators of beliefs and attitudes than traditional training methods. They can pass the message across clearly and effectively to workers.

Instilling fear can help a lot to improve safety. There are only two ways to make workers change their behaviours; either financially or showing them that they are going to lose everything. Most training programmes available in the industry function like a matter of ticking a box and they do not instil fear. However, the system presented will be very useful and could satisfy the need of most contractors.

Visual impacts can convey the message very effectively in that watching workers falling in real construction sites and suffering permanently as a consequence can help workers change their attitude towards safety and discourage them from taking short cuts.

Utilisation of the system

Respondents indicated several possible ways that the proposed system could be used to bring about attitudinal change. The comments made by the respondents are as follows:

The proposed system can be used in a range of ways, in tool box and pre-start talks and in trade or supervisor education and training programmes in technical institutions.

The web-based tool can be used for providing activity-based safety training for workers and less-experienced/competent supervisors. It is a very good tool for site-specific inductions that are done just before starting an activity.

The system will be useful for contractors, subcontractors and supervisors as a refresher training tool.

It can be used on remote sites irrespective of their locations. When the system is implemented on a main contractor's server, subcontractors and labour subcontractors can be provided with access privileges and they can use it for more effective inductions.

The system can be helpful to show videos at tool box talks and pre-start meetings and inform workers of possible dangers. Upon showing the videos, signed acknowledgement letters could be obtained from workers saying 'I watched the video and understand the safety precautions necessary for the activity'. This will save contractors from lawsuits and make workers more responsible for their actions.

Having a laptop and mobile internet connection will enable any supervisor or subcontractor to use this system anywhere on a construction site.

Benefits of the system

Survey respondents indicated many potential benefits of utilising the proposed system as quoted below.

There are benefits for all parties involved in onsite construction. For workers – they can go home fully every day. For supervisors, managers and employers – it could provide a drive for the implementation of their own OHS systems. Many organisations have some sort of OHS system in place but they may not be implemented in practice. The proposed system would reinforce their implementation.

Conducting safety inductions will be easy with the system and it is also a good educational tool.

In the long run the system can help to save money for contractors and subcontractors in that insurance premiums and overhead costs related to accidents will be reduced.

The visual resources will be very helpful to improve the OHS culture of the industry.

The videos and the contents explicitly explain to everyone what their roles and responsibilities are. Given the cohort of workers who are vastly secondary school qualified, these videos and simple explanations would be easily understandable. Moreover, short videos are very helpful in that workers do not like to participate in long training sessions as they are largely time and wage conscious.

Having a web-based database is very useful for contractors who have projects on the outskirts of the state. Though the sites are remote and difficult to monitor, when an accident happens its impact shakes the heartland directly. Systems like this can be very useful for proving training to workers in a company.

The system can help to curtail accidents and project downtime, improve OHS compliance, awareness and understanding and save time and cost for workers and contractors.

Harnessing its utilisation

Many suggestions were offered by respondents to harness the utilisation of the proposed system on site as well as to improve the absorption level of the message by workers.

Industry-used terminologies should be used instead of academic terminologies as the main audience is workers. Having simple, short and direct to the point and trade-specific videos will help to get to all levels of people.

Showing a balanced story is important. Showing only frightening videos may de-motivate people to work in the industry. So every video resource should be a balanced piece.

Because the system is a powerful tool and can contribute to accident prevention, the system should be made available to all workers. The construction industry is characterised by having a significant portion of workers who are of non-English-speaking backgrounds. In order to reap better outcomes, the system should contain multilingual videos.

It would be better to provide links to external resources such as best practice manuals, codes of practice and other WorkCover publications that are related to the video contents in the proposed system.

Most of their suggestions have already been absorbed into the new training system. However, the suggestions related to multilingual resources and links to external resources were not addressed at this stage of the research.

Obstacles for utilisation

There was a strong feeling amongst participants that there is high demand for OHS training resources like this in the industry and therefore the implementation of this will not face major challenges. Nevertheless, they also identified possible managerial and technological obstacles to the successful implementation of the proposed system in the industry as quoted below.

> Small builders and subcontractors may not have access to computers and internet connections in the workplace.

> There may be a percentage of site supervisors and managers who are not computer literates, have fear of technology or/and do not realise the benefits the system could provide. They tend to believe more in the traditional mode than innovative approaches. This could eventually turn into resistance to adopting the proposed system.

> The internet connection quality in very remote sites is quite inadequate for streaming videos, which may cause an obstacle.

> Updating videos in pace with legislative changes may be costly for builders.

The findings of the evaluation exercise suggest that the proposed training scheme, together with its associated web-based system, is a successful endeavour. It also validates the hypothesis of this research that the utilisation of training resources, which are deliberately designed to simultaneously stimulate anxiety or fear of death, incapacitation and/or injuries in workers while delivering learning contents, could result in better concentration and cognition by workers. This would eventually result in permanent safety consciousness in the minds of workers that can influence their behaviours on site and thereby minimise accidents.

Conclusions

Unsafe behaviours of workers have been repetitive causes of accidents in construction. Current OHS training schemes that are used by the construction industry seem to have failed in dealing with this issue successfully. This research developed the affective OHS training scheme, as a complement to the existing training regimes, to better deal with attitudes of workers towards safety and thereby alter their behaviours positively. A web-based system was developed to provide interactive training resources, which facilitate affective OHS training to construction workers in site inductions and other OHS training programmes. The new web-based tool was evaluated by its potential end-users from the construction industry and the findings suggest that the tool have many positive implications for construction workers and builders, including the following.

1   Constant utilisation of interactive learning resources in the e-tool in site inductions, toolbox talks and other OHS training programmes could change even hardened negative attitudes in workers.

2   The e-tool can help builders to establish proofs of providing OHS training to workers on their sites, irrespective of their locations and remoteness.

Nonetheless, as always expected in any information system implementation, there may be some initial managerial and technological hurdles that organisations need to overcome to get this training scheme up and running smoothly. Overall, the study findings suggest that the implementation of the proposed system in the construction industry would help address the challenges posed by unsafe behaviour as root causes of accidents and thereby minimise accidents on site.

Further studies are suggested to investigate how similar affective training tools may be developed for other hazardous trades such as excavation works, crane and lifting devices use, mobile plant use, and hand and power tool use.

4 E-OHS monitoring system for construction projects

Introduction

The successful monitoring and tracking of the implementation of OHS management systems on construction sites is a critical step to provide a healthy and safe work environment and to prevent accidents (Wilson and Koehn 2000; Tam *et al*. 2001; Hinze and Gambatese 2003). Deployment of an effective OHS monitoring and tracking system is therefore crucial for builders. Nonetheless, there are many critical issues in construction that challenge effective deployment of OHS monitoring and tracking systems. Among the challenges are:

- the constantly changing hazards as the project is constructed;
- multilayered subcontracting and the loss of track of OHS responsibility distribution;
- concurrent progress of multiple tasks on site;
- abundance of hazardous material use, plant operations, tradesmen and subcontractors;
- limitations of supervisors to monitor every aspect meticulously;
- large amount of paperwork and the risk of losing data/documents; and
- active involvement of top management in monitoring OHS performance on site becomes limited without real-time summaries of performance.

However, the capabilities that information and communication technologies (ICT) provide can assist builders in overcoming these challenges and improving the OHS monitoring process. Development of ICT tools for OHS monitoring and tracking has attracted only a marginal level of interest amongst researchers. Some previous developments of computer systems deal with a couple of aspects of OHS monitoring and tracking. Chua and Goh (2004) and Cheung *et al*. (2004) developed incident information systems to provide assistance with collection and analysis of incident data for: (1) summarising accident rates, trends and patterns, (2) identifying and investigating accident causes and (3) estimating the costs of accidents. Research efforts for developing OHS audit tools are also noted. BRE, UK developed a construction OHS audit tool called SABRE (Lingard and Rowlinson, 2005). It is evident that the development of a comprehensive OHS monitoring and tracking system for construction is still a potential area for exploration. Moreover, collaborative project management using project extranets has reached a reasonably mature state in construction but its extension into OHS management is still at a primitive stage. This chapter discusses the development of a groupware system for OHS monitoring in construction. First, a new framework for collaborative OHS monitoring using a platform web is presented. Secondly, the findings of system analysis for the proposed system are discussed, followed by system development. Finally, an initial system evaluation and its findings are elaborated, followed by conclusions.

P₃O framework for collaborative OHS monitoring

The prime objectives of OHS monitoring and tracking are demonstrating OHS compliance and continual improvement of OHS on site. The P_3O framework is proposed, as illustrated in Figure 4.1, as a means to assist builders in achieving these. The framework endeavours to achieve the objectives by tightly integrating OHS monitoring processes with the project team through a collaborative platform that embodies OHS action protocols.

   Construction OHS management is the responsibility of everybody in a project team, which means successful implementation of OHS systems on site requires the involvement of many project stakeholders, including the site team, OHS officer, top management of the builder and client's representative/ PM. However, such a collaborative effort may not be possible in the absence of an ICT infrastructure with appropriate collaborative procedures and proformas embedded. With the support of the ICT infrastructure and active involvements of project stakeholders, OHS implementation data can be captured regularly from sites and OHS performance indicators can be produced. These indicators could help to identify collaborative corrective actions from project team members. The ICT infrastructure can not only help traverse through this process towards continual OHS improvement

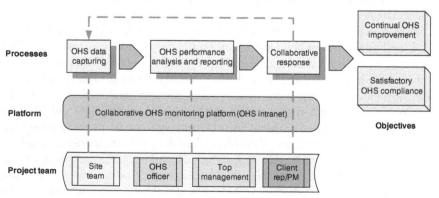

Figure 4.1 P₃O framework for collaborative OHS monitoring

on site, but can simultaneously record and demonstrate satisfactory OHS compliance on site by builders.

Given the dispersed nature of construction projects and project team members and the myriad of OHS data, a web-based platform, driven by a sound OHS database, is paramount for the practical application of the P_3O framework. Like any information system, the development of the web-based OHS monitoring system also entails the adoption of a system approach for a successful system delivery. The system approach involves a step-by-step process to system delivery, including system analysis, system design/modelling, system implementation/prototyping and system testing and evaluation. Each stage in the process is essential and involves varying tactics. The ensuing sections describe these steps in the context of the proposed web-based OHS monitoring system.

System analysis for OHS monitoring system

Comprehensive literature reviews and contents analyses were carried out to explore and understand the detailed cross-section of OHS implementation, monitoring and record-keeping requirements for construction projects. These detailed reviews and analyses studied OHS legislations, various OHS codes of practice, best practice guides, standards, protocols and frameworks and research publications to understand: (1) what tasks are involved in managing safety on construction sites, (2) who are involved in each task and what their roles are, and (3) what data need to be kept for compliance requirements as well as for internal reviews and improvements. These helped to develop procedures, protocols and templates for OHS implementation, monitoring and record-keeping. In summary, the following components are required in order to implement effective OHS management systems on site:

- delegating OHS roles and responsibilities to project team members;
- implementing safe work method procedures that were documented in safe work method statements (SWMSs), and continually reviewing hazards and revising control measures;
- conducting regular safety consultations, tool box talks and safety inductions with workers;
- administering a plant and machinery inspection subsystem;
- implementing an electrical equipment safety inspection subsystem;
- maintaining a hazardous substances safety subsystem;
- implementing a permit-to-work subsystem;
- managing a regular OHS audit subsystem;
- maintaining an accident reporting and investigation subsystem.

The sections below elaborate on each of these components and derive procedures, protocols and templates for their implementation, monitoring and record-keeping.

OHS responsibilities and project team members

Safety is everybody's responsibility. In order to create a safe and productive workplace all members of a project team should take a stake and responsibility for implementing the OHS system on site. However, construction projects are complex and characterised by multilayered subcontracting. Many hazardous tasks occur concurrently to meet tight schedules. A critical issue for OHS system implementation in construction projects is therefore the clear allocation of OHS roles and responsibilities to different subcontractors and people involved in the project. Besides sharing OHS responsibilities at a site level, management commitment is one of the most important factors at an organisational level. Studies show that support from management is crucial for improvement in construction safety (Ng *et al.* 2005). Hence, a clear definition of OHS roles and responsibilities of subcontractors and individuals such as senior management, general manager, project manager, site manager, site supervisors/foreman and OHS officer should be maintained (Lingard and Rowlinson 2005). This should indicate who should do what, to what standard and when, in relation to the management of OHS. For example, supervisors or foremen could be expected to conduct weekly tool box meetings with workers, a site manager could be expected to induct all new workers and visitors who enter the site and a project manager could be expected to examine the past OHS performance of all subcontractors. Additionally, it is equally crucial to ensure that employees are adequately trained in OHS to carry out the roles and responsibilities assigned to them.

Two forms have been developed, as shown in Exhibit 1 and Exhibit 2, for record-keeping of OHS roles and responsibilities delegations to project team members. While Exhibit 1 helps to record basic details about a project, Exhibit 2 records individual employees' OHS responsibilities, competencies and OHS task assignments from time to time.

OHS committee and consultations

Consultation with workers on OHS matters has been encouraged in order to provide an agreed mechanism where employees can discuss and express their views. Employees have valuable information on how the job is done on site because they are closer to the work being carried out on a daily basis. Therefore, their solutions to problems could improve the OHS management system (SafetyNet 2010). Workers' involvement in OHS is a critical requirement of legislation, therefore the establishment of a consultative mechanism is an important part in organising for OHS. An efficient way of demonstrating worker participation in OHS is to form a joint employer–employee OHS committee. WorkSafeBC (2009a) defined a joint OHS committee as 'a committee made up of worker and employer representatives working together to identify and resolve health and safety problems in the workplace'. While the employer is responsible for the overall

Project Information					
Project Details					
Project Id/Contract Id					
Description of work					
Client					
Project duration	Start :		Expected completion:		
Head Contractor Details					
Business name					
Address					
Director/Manager					
Contact no	Phone:		Mobile		
Fax					
Email					
Project OHS Officer					
Name					
Contact no	Phone:		Mobile		
Fax					
Email					
Subcontractor Details					
Business name	Address	Contact person	Phone no	Mobile no	Email

Exhibit 1 Project information form

Employee details			
Full name			
Position			
Contact details	Email:	Mobile:	
Supervisor			
Emergency contact	Name:	Contact no:	
OHS training/competency details			
Training received	Card No. / Reg. No.	Date received	Refresher date
OHS responsibilities			
Task assignment			
Task assigned	Date	Due date	Progress

Exhibit 2 Employee OHS responsibility form
Source: Adapted from CSA (2008)

health and safety programme, the committee is encouraged to identify and recommend solutions to problems. In the context of the construction industry, it is important that subcontractors and their workers are involved in the consultation process (Lingard and Rowlinson 2005).

An OHS committee meeting is a means of providing employer and worker representatives with positive participation and cooperation in promoting health and safety at work. Meeting reports should be objective and brief,

OHS Committee Meeting Invitation		
Project		
Meeting no		
Invitation by	Full name:	Job title:
Date & time of meeting	Date:	Time:
Meeting venue		
Message to attendees		
Invitees		
Full name	**Job title**	**Employer/Subcontractor**
Agenda items		
Attachments (if any)		

Exhibit 3 OHS committee meeting
Source: Modified from CSA (2008), Mel Crook & Associates (2010), SafetyNet (2010)

OHS Committee Meeting Minute							
Project							
Meeting no							
Date & time of meeting	Date:		Time:				
Meeting venue							
Attendees							
Apologies							
Actions & Agreements							
Item #	Agenda Item	Discussion summary	Action required	By who	By when	Action completion date	Attachment (if any)

Exhibit 4 OHS committee meeting minute
Source: Modified from CSA (2008), Mel Crook & Associates (2010), SafetyNet (2010)

providing a historical record and noting decisions and actions that have been recommended by the committee. It should list actions required and taken by whom and deadlines (WorkSafe BC 2009a). Accordingly, two forms have been developed to facilitate the OHS committee meeting and consultation process as shown in Exhibit 3 and Exhibit 4.

Worker registration and personal protective equipment use

Workers on construction sites should be protected from hazards and poor work conditions. The best way of doing this is by using engineering or work practice controls, for example, installing a barrier between the hazard and workers or changing work processes. However, when these controls are not possible or adequate, personal protective equipment should be provided for workers to ensure their safety (OSHA 2003). Personal protective equipment, commonly referred to as 'PPE', is equipment to minimise exposure to

Details of worker							
Full name			Worker's registration No.				
Address			Induction card No.				
Trade			Employer/subcontractor				
Emergency contact	Name:		Relationship:			Contact no:	
Medical data	Any regular medication:			Any allergies:			
Pre-start tests & PPE							
Date				By whom			
Test	Pass		Fail		PPE issued		
Alcohol/drug test					Hard hat	Respiratory equipment	
Competence test							
Work activity involved					Safety footwear	Ear protection	
					Eye protection	Sun glasses	
					Safety harness	Sun screen	
					Overall		
					Illuminating safety vest		
Additional Comments:							
Signature of worker	*I have received the listed PPE with appropriate instruction/training in its correct use.*						

Exhibit 5 PPE record form
Source: Evolved from OSHA (2003), CSA (2008), Mel Crook & Associates (2010), SafetyNet (2010)

hazards, which include hardhat, safety footwear and eye and hearing protection devices. PPE required for a specific construction task is specified in the safe work method statement (SWMS) for the task. For the greatest possible protection for workers, organisations should ensure that all items of PPE are manufactured, used and maintained in accordance with relevant standard (CSA 2008). Also, workers must be instructed or trained in the correct use of each PPE prior to use.

A form can be used to record all PPE supplied to workers as specified, as a control measure, in the SWMS for a given construction task. In addition to the PPE issued details, workers' particulars and pre-start check details can also be recorded for greater control. A form has been developed for this purpose, as shown in Exhibit 5, by a content analysis of materials from OSHA (2003), CSA (2008), Mel Crook & Associates (2010) and SafetyNet (2010).

Continual hazard monitoring and review

A safe work method statement (SWMS) is required for each activity carried out on site. This statement lists hazards and potential risks associated with a construction activity and control measures required to avert the risks. SWMSs can serve to prove that a contractor has implemented necessary measures to make the workplace safe. However, the construction process is dynamic and the nature of hazards may change during its course. It is therefore crucial that continual hazard monitoring and reporting is practised to prevent incidents. All workers, employees and subcontractors are encouraged to report hazards

Hazard Report			
Person reporting the hazard to complete			
General			
Date			
Workplace/Project			
Submitted By	Name:	Position:	
Submitted To	Name:	Position:	
Details of Hazard/Risk			
Work Activity			
Location of Hazard			
Hazard description			
Attachments			
Risk Class	High (1) ☐	Medium (2) ☐ Low (3) ☐	
Control Measures			
Manager responsible to complete			
Corrective action required			
By whom		When	Immediate
Action completion			
Employee responsible to complete			
Action taken		By whom	
Date		Time	
Manager responsible to complete			
Approval	☐ Approved ☐ Not approved	By whom	

Source: Adapted from CSA (2008), SafetyNet (2010).

Exhibit 6 Hazard reporting form
Source: Adapted from CSA (2008), SafetyNet (2010)

immediately to their work supervisor, taking actions to minimise any harm caused. Upon a new hazard being reported, the site management team must investigate it and take appropriate control measures to eliminate or minimise risks.

A form has been developed, as shown in Exhibit 6, to facilitate the hazard reporting and response process based on a contents analysis of two documents. The form has three parts to help with the hazard reporting and response process: (1) the first part is completed by the person reporting a hazard and sent to the manager responsible for the construction site; (2) the second part is completed by the manager responsible who identifies corrective actions and assigns employees to implement the corrective action; and (3) once the actions have been attended to, the employee responsible will record the actions taken and this is confirmed by the manager responsible. This concludes the hazards reporting and response process.

Tool box talks and safety inductions

A tool box talk, pre-start talk, safety chat or tailgate meeting is a brief safety talk or meeting at the beginning of a shift. These talks are typically brief (two to five minute) interactive discussion meetings. They are used to cover a variety of short safety training subjects and to remind workers of the importance of acting safely each day before they start work. Tool box talks also enable supervisors to check out whether safety awareness is maintained throughout the project. Under the OHS legislation, contractors are required to provide site- and activity-specific safety inductions to workers before commencing work. Toolbox talks can be viewed as a suitable opportunity for that. The site induction is mandatory for all workers including subcontractors prior to starting work on site to be aware of OHS issues. It provides participants with knowledge of the contractor's rules and procedures for site safety, emergency management, supervisory and reporting arrangements and other site-specific issues on a particular construction site (Australia Safety and Compensation Council 2007). The task-specific induction provides information and instructions to anyone undertaking a particular construction activity of the risk factors and control measures relating to the task. It should be conducted prior to commencing high-risk construction work, following the safe work method statement (SWMS) for the task. The topics of task-specific induction include common hazards, risks and control measures, relevant legal responsibilities, codes of practice or standards and safe work methods to be used (Australia Safety and Compensation Council 2007).

All tool box talks should be recorded on a form, addressing specific OHS issues on site. Any points can be raised and all corrective actions recorded on the form should be implemented by the nominated person. Site supervisors are required to ensure that all corrective actions are completed and reviewed for effectiveness (CSA 2008; WorkSafeBC 2010). A form has been developed as shown in Exhibit 7 for recording tool box talks by a contents analysis of documents produced by CSA (2008), Australia Safety and Compensation Council (2007), Mel Crook & Associates (2010) and SafetyNet (2010). This form has four parts: (1) general details, (2) attendees details, (3) topics discussed and (4) corrective actions required. During a pre-start talk session, the presenter needs to record the general details, including attendees and the topic discussed in relation to hazards in the construction activity to be performed. If any corrective actions are identified in the pre-start talk session, a request will need to be made by the presenter and a person will be nominated for action. Once the actions are attended to, the nominated person will record the action taken against actions requested.

Hazardous substances safety

A hazardous substance can be any substance, liquid, solid, dust or gas that may cause harm to workers. Examples of hazardous substances used on construction sites include asbestos, fibreglass, synthetic mineral fibres, cement dust, exhaust fumes, glue, paint and solvents. Hazardous substances should not be a problem when things are not going wrong (Department of Commerce 2010). In order to monitor their correct use on site, a hazardous substances register may be used. The hazardous substances register comprises a list of all the hazardous substances used and stored on site, their locations

Tool Box Talk Form					
Workplace:					
Work activity:					
Tool box talk reference:					
Supervisor/Presenter:					
Date :	Start time:			End time:	
Attendees					
Full name		*Signature*			
Topics/Issues Discussed					
Site induction issues discussed:					
Task-specific induction issues discussed:					
Other issues discussed:					
Attachments (if any)					
Corrective Actions					
Tool box talk presenter to complete			*Employee responsible to complete*		
Action required	*By whom*	*Due*	*Action taken*	*Date*	*Remarks*
		Immediate	☐ Complete		
		Within 24 hrs	☐ Incomplete		
		Within 7 Days			

Exhibit 7 Tool Box Talk Form
Source: Adapted from Australian Safety and Compensation Council (2007), CSA (2008), Mel Crook & Associates (2010), SafetyNet (2010)

and material safety data sheets (MSDS) for all those hazardous substances. The register also has to contain a notation against each hazardous substance as to whether a risk assessment has been completed and made available to all workers potentially exposed to the hazardous substances (Department of Commerce 2010). The register can allow the generation of a list of hazardous substances that have not been risk-assessed. Exhibit 8 shows a simple template for the hazardous substances register.

Plant safety

In relation to plant and machinery, there are risks associated with mechanical hazards, non-mechanical hazards and access hazards. Mechanical hazards are related to machine moving parts which may have sufficient force in motion to cause injuries to workers. Non-mechanical hazards include chemicals, harmful emissions, fluids, gas, electricity and noise, which can cause injuries or harm if not properly controlled. Workers must be provided with safe access around plant and machinery to prevent any risk of fall or injury (Department of Consumer and Employee Protection 2010). In order to reduce risks, plant and machinery should be inspected and maintained in accordance with relevant standards and manufacturers' recommendations on a regular basis (CSA 2008).

Hazardous Substances Register	
Substance name	
Product labelled	☐ Yes ☐ No
Hazardous	☐ Yes ☐ No
Quantity on site	
Location/process where it is used	
Risk assessment completed	☐ Yes ☐ No
Attach MSDS	
MSDS issue date	
Remarks	

Exhibit 8 Hazardous substances register
Source: Adapted from CSA (2008), SafetyNet (2010)

Three types of inspection and maintenance schemes are suggested in literatures related to OHS, including regular service and maintenance of plant at its due date, pre-start check of plant and periodic plant inspection. Three templates have been developed for this purpose, as explained below.

Plant register

This holds general details of plant that are being used on site. The details include plant description, registration details and maintenance details. Keeping these records will enable the generation of a list of plant whose maintenance and/or reregistration is overdue and/or falling soon. Exhibit 9 shows a plant register that has been developed for this purpose.

Pre-start checklist

This allows recording of pre-start check details for selected plant in the plant registry. Pre-start OHS risk assessments for plant cover three components: (1) walk-around inspection of the plant and its parts, (2) the site and working environment where the plant is operated and (3) safe access and plant movement. A checklist has been developed for this as illustrated in Exhibit 10.

Plant Register		
Plant		
Serial no./registration no.		
Make/model		
Plant provider		
Registration expiry		
Date of last service		
Required service frequency		
Next service due	**Date:**	**Other frequency:**
Period on site		
Log book available (Y/N)		

Exhibit 9 Plant register
Source: Modified from CSA (2008), SafetyNet (2010)

Plant Pre-start Check			
Plant			
Serial no.			
Date			
Operator's name			
Plant operator's certificate/ticket no.	Expiry date		
Induction card no.			
Plant inspection by			
Inspection verified by			
Plant			
1. Is visibility adequate? (windscreen, windows, wipes, washers, mirrors)	☐ Yes	☐ No	☐ NA
2. Is the cabin safe? (access, egress, seatbelts, plant control labels)	☐ Yes	☐ No	☐ NA
3. Are warning devices working (flashing lights, amber beacon, horn, reverse alarms)	☐ Yes	☐ No	☐ NA
4. Is the brake in good condition? (emergency and service)	☐ Yes	☐ No	☐ NA
5. Are the fire extinguishers serviceable & accessible?	☐ Yes	☐ No	☐ NA
6. Are all the parts safe to operate? (chains, lift booms & hooks, tyres/tracks/wheels)	☐ Yes	☐ No	☐ NA
7. Are the levels of fluids adequate? (transmission oil, hydraulic oil, engine oil, brake fluid, power steering fluid, coolant & fuel etc)	☐ Yes	☐ No	☐ NA
8. Are the service records up to date?	☐ Yes	☐ No	☐ NA
9. Are the operator's manual and load charts are in cabin?	☐ Yes	☐ No	☐ NA
10. Is the operator certified/licensed to operate the plant?	☐ Yes	☐ No	☐ NA
11. Is a hazard assessment checklist available for the plant?	☐ Yes	☐ No	☐ NA
Site and working environment			
Are adequate control measures taken to protect the operator and workers in the area from risks due to:			
1. Powerlines	☐ Yes	☐ No	☐ NA
2. Trees & vegetation (combustible)	☐ Yes	☐ No	☐ NA
3. Overhead service lines	☐ Yes	☐ No	☐ NA
4. Bridges	☐ Yes	☐ No	☐ NA
5. Water ways, dams or tidal activities	☐ Yes	☐ No	☐ NA
6. Surrounding buildings	☐ Yes	☐ No	☐ NA
7. Other equipment in the area	☐ Yes	☐ No	☐ NA
8. Dangerous substances	☐ Yes	☐ No	☐ NA
9. Underground services	☐ Yes	☐ No	☐ NA
10. Recently filled trenches/excavations	☐ Yes	☐ No	☐ NA
11. Temperature from the plant	☐ Yes	☐ No	☐ NA
12. Noise/vibration from the plant	☐ Yes	☐ No	☐ NA
13. Lighting level	☐ Yes	☐ No	☐ NA
14. Uneven/unstable/slippery/unprotected floor around the plant	☐ Yes	☐ No	☐ NA
15. Crushing/strikes	☐ Yes	☐ No	☐ NA
Safe access and machine movement			
1. Is there safe access to the work area?	☐ Yes	☐ No	☐ NA
2. Is there room for movement of plant equipment on site?	☐ Yes	☐ No	☐ NA
3. Are the access roads/haul roads clear and serviceable?	☐ Yes	☐ No	☐ NA
4. Are there light vehicle movements on site?	☐ Yes	☐ No	☐ NA
5. Is traffic control adequately?	☐ Yes	☐ No	☐ NA
Remarks (if any):			

Exhibit 10 Pre-start checklist
Source: Adapted from WorkSafe BC (2009b)

Periodic Plant Inspection				
General details				
Plant description				
Plant no.		Serial no:		
Plant provider				
Inspection by		Date		
Inspection report submitted to (manager)				
Inspection items		**Check**		
1. Risk assessment for plant operation documented		Yes ☐	No ☐	NA ☐
2. Operator's manual available		Yes ☐	No ☐	NA ☐
3. Maintenance reports available		Yes ☐	No ☐	NA ☐
4. Log book is maintained		Yes ☐	No ☐	NA ☐
5. Certification/competency of operator acceptable		Yes ☐	No ☐	NA ☐
6. Regulatory authority registration is current		Yes ☐	No ☐	NA ☐
7. Fire extinguishers serviceable & accessible		Yes ☐	No ☐	NA ☐
8. Crack test reports acceptable		Yes ☐	No ☐	NA ☐
9. Chains tested and tagged		Yes ☐	No ☐	NA ☐
10. Tested and tagged electrically		Yes ☐	No ☐	NA ☐
11. Roll over protection (ROPS) provided		Yes ☐	No ☐	NA ☐
12. Seat belts available		Yes ☐	No ☐	NA ☐
13. Brakes in good order (emergency and service)		Yes ☐	No ☐	NA ☐
14. Warning devices work (flashing lights, amber beacon, horn, reverse alarms)		Yes ☐	No ☐	NA ☐
15. Visibility adequate (windscreen, windows, wipes, washers, mirrors)		Yes ☐	No ☐	NA ☐
16. Cabin is safe for the operator (access, egress, temperature, vibration/noise, lighting, ergonomics)		Yes ☐	No ☐	NA ☐
Remarks				
Control Measures				
Manager responsible to complete				
Corrective actions required				
By whom		Due date		
Action completion				
Plant provider to complete				
Action taken				
Date				

Exhibit 11 Periodic plant inspection form
Source: Modified from CSA (2008)

Periodic plant inspection

Exhibit 11 shows a template that facilitates periodic plant inspections. The template allows recording of regular plant inspection results for selected plant in the plant register and informing the manager responsible/plant provider of any corrective actions required.

Electrical equipment safety

All electrical equipment including leads, portable power tools, junction boxes and earth leakage or residual current devices should be inspected and tested by a qualified person and labelled with a tag of currency prior to being

Electrical equipment register	
Equipment ID/serial no.	
Equipment description	
Date of test	
Test result	
Date of next test	
Tested by	
Licence/registration no.	

Exhibit 12 Electrical equipment register

used on the construction site. An electrical equipment register can be used to list all electrical equipment brought on site and record test details prior to commencing work. It also needs to be maintained during the construction on site (CSA 2008). While this will satisfy record-keeping, it will also facilitate easy identification of equipment that need refresher tests. Exhibit 12 shows a sample electrical equipment register.

Permit-to-work system

A permit-to-work (PTW) system is a formal written system used to control certain types of work that are potentially hazardous. It is a document which specifies the work to be done and the precautions to be taken (HSE 2010a). Permit-to-work systems are required to manage certain high-risk activities and prevent major accidents on construction sites. Activities requiring a permit to work include deep excavations, maintenance works on chemical plant, roof access, works in confined spaces, high voltage electricity entry, hot work and activities categorised as high risk. A permit-to-work allows work to start only after safe procedures have been defined and it provides a clear record that all predictable hazards have been considered.

A template has been developed, as shown in Exhibit 13, to facilitate the PTW process, which contains three parts, including: (1) a part for site personnel to initiate requests for PTW from an approving person, (2) a part for approving the PTW and (3) a part for PTW cancellation.

OHS audit system

An OHS audit is a structured process of collecting independent information on the efficiency, effectiveness and reliability of an OHS management system and corrective action plans (Teo and Ling 2006). The OHS audit could measure the success of the implementation of OHS systems and encourage positive achievement (Ahmad and Gibb 2003). All aspects of the OHS management process should be subject to regular systematic evaluation through OHS audits that reveal how well an OHS management system is functioning. Audits present valuable opportunities for managers to identify weaknesses in the OHS management system and to develop methods to improve the effectiveness of the system (Lingard and Rowlinson 2005).

Permit-to-work form					
Permit-to-work request					
General details					
PTW no.					
PTW for task					
Project/site					
Location of work					
Permit valid	From Date:		Time: hrs	To Date:	Time: hrs
PTW requested by	Name			Position	
Date of request					
PTW requested from	Name			Position	
Contractor in-charge of work					
Responsible site person					
Description of work					
Description of hazards/risks identified					
Control measures/precautions taken					
Isolation of hazards					
Atmosphere checks					
Engineering controls					
Administrative controls					
Personal protective equipment					
Other controls/precautions					
Emergency response plan					
Standby personnel					
Worker sign on/off					
I have been advised of and understand the control measures and precautions to be observed during the work.					

Date	*Name of worker*	*Time on*	*Time off*	*Signature*

Permit-to-work approval				
Inspections	All hazards of the operation are identified		☐ Yes ☐ No	
	Control measures and precautions are appropriate for the operation		☐ Yes ☐ No	
PTW	☐ Approved ☐ Disapproved			
Remarks/conditions of PTW				
Date				
PTW valid	From Date: Time: hrs	To Date: Time: hrs		
Authorised person	Name:	Position:	Signature:	
Notification to				
Permit-to-Work Cancellation				
Date of cancellation				
Reasons				
Authorised person	Name:	Position:	Signature:	
Notification to				
Attachments				

Exhibit 13 Permit-to-work template
Source: Modified from UWA (2009), HSE (2010a), Port of Brisbane (2010)

Charles *et al.* (2007) further reinforced that audits provide a systematic evaluation of project safety performance, enable the timely identification and rectification of safety problems and provide a basis for feedback to the constructor regarding safety performance. In an audit, the entire OHS management system is evaluated, which includes the auditing of the OHS policy, performance standards, planning process, organisational arrangement and measurements and monitoring activities.

Audits are often undertaken by third-party auditors based on a predetermined set of measurement items. However, cross audits, done by construction personnel from other projects of the same builder, are also valuable. The cross audits provide an independent review of OHS management processes and also give a mechanism for sharing best OHS practices between projects, thus strengthening OHS culture organisation-wide (Charles *et al.* 2007). The results of audits must be clearly documented and communicated to the constructor, which can lead to improvements in OHS management implementation. A comprehensive template has been developed, as illustrated in Exhibit 14, to facilitate this crucial process of OHS management.

Incident management

The discussions so far have focused mainly around prevention of occupational injuries and illness. Despite effective prevention measures being in place, the construction industry is still recording dismal accident rates all over the world. It is therefore equally crucial to implement a sound incident managing subsystem on site to minimise harm to both individuals and the organisation. It is also a legal obligation for an organisation to act responsibly and reasonably to minimise harm to people in the event of an incident.

An incident can be defined as any unexpected event that occurs as a result of work (or any activity undertaken on site), and that results in or has the potential to cause injury, illness or other loss (Lingard and Rowlinson 2005). Incidents not only include injuries, illness and damage but also near-misses that have the potential to harm. A sound incident management entails a sequence of responses and actions from a builder in the event of an incident, including the following.

- Emergency management at the incident scene to ensure people are safe and the hazard is removed or isolated to minimise its impact on people. This includes evacuating the site and erecting barricades or cordoning off incident areas, which also ensures evidence is preserved for incident investigation.
- Providing prompt first-aid treatment to injured workers to reduce the severity of injuries.
- Reporting the incident, no matter how minor it is. There are two types of incident reporting, internal and external. All incidents should

Project details				
Project				
OHS audit no				
Date of audit				
Audit team members				
Report sent to				

OHS audit checklist				
OHS audit team to complete			Site management to complete	
Audit item	Confirm	Suggested corrective actions	Action taken	Date
General/OHS management Items				
1. Project details/description of works/ organisation details are current	Y ☐ N ☐ n/a ☐			
2. OHS policy signed and dated by director/manager	Y ☐ N ☐ n/a ☐			
3. OHS roles and responsibilities are allocated and signed	Y ☐ N ☐ n/a ☐			
4. Hazards are identified and risks are assessed continually	Y ☐ N ☐ n/a ☐			
5. Hazard reports are completed for continual risk assessment	Y ☐ N ☐ n/a ☐			
6. Controls for high-risk activities are documented in Safe Work Method Statements	Y ☐ N ☐ n/a ☐			
7. Training and competency register is current	Y ☐ N ☐ n/a ☐			
8. Site-specific induction training records are current	Y ☐ N ☐ n/a ☐			
9. SWMS training is current	Y ☐ N ☐ n/a ☐			
10. Consultation and tool box talk arrangements (nature, topics, intervals) are documented	Y ☐ N ☐ n/a ☐			
11. Plant/equipment register is current	Y ☐ N ☐ n/a ☐			
12. Hazardous substances/dangerous goods register is current	Y ☐ N ☐ n/a ☐			
13. Personal protective equipment register is current	Y ☐ N ☐ n/a ☐			
14. Electrical equipment register is current	Y ☐ N ☐ n/a ☐			
15. Periodic workplace inspection checklists are completed	Y ☐ N ☐ n/a ☐			
16. Register of Injuries is current	Y ☐ N ☐ n/a ☐			
17. Incident investigation reports are completed	Y ☐ N ☐ n/a ☐			
18. Injury management and return-to-work programme is displayed	Y ☐ N ☐ n/a ☐			
19. Workers compensation information is current	Y ☐ N ☐ n/a ☐			
20. Safety meetings held periodically	Y ☐ N ☐ n/a ☐			
21. Changes and distribution of the OHS Mgt Plan are recorded	Y ☐ N ☐ n/a ☐			
Site and housekeeping				
22. Posters and safety signs/warning in place	Y ☐ N ☐ n/a ☐			
23. Access paths defined (signage tape, other)	Y ☐ N ☐ n/a ☐			
24. Access paths kept clear	Y ☐ N ☐ n/a ☐			
25. Everyone has clear access and exit	Y ☐ N ☐ n/a ☐			
26. Open ditches protected	Y ☐ N ☐ n/a ☐			
27. Equipment secured	Y ☐ N ☐ n/a ☐			

Exhibit 14 OHS audit template
Source: Adapted from WorkSafe BC (2009b), CSA (2008), HSE (2010a)

continued ...

Exhibit 14 continued

28. Drop-offs protected	Y☐ N☐ n/a☐			
29. Nails or sharp objects removed or bent down	Y☐ N☐ n/a☐			
30. Ladders lowered	Y☐ N☐ n/a☐			
31. Leads suspended	Y☐ N☐ n/a☐			
32. Floors free from tripping hazards	Y☐ N☐ n/a☐			
33. Materials stacked	Y☐ N☐ n/a☐			
34. Work area lit	Y☐ N☐ n/a☐			
35. Bins available and in use	Y☐ N☐ n/a☐			
36. Washrooms and eating areas kept clean	Y☐ N☐ n/a☐			
37. Free of refuse, scrap, overgrown vegetation	Y☐ N☐ n/a☐			
38. Fencing around drip line of retained trees				
39. No material/equipment stored within drip lines	Y☐ N☐ n/a☐ Y☐ N☐ n/a☐			
Emergency/fire/injury/first aid				
40. Alarm system, serviceable	Y☐ N☐ n/a☐			
41. Evacuation procedure in place	Y☐ N☐ n/a☐			
42. Emergency contacts displayed	Y☐ N☐ n/a☐			
43. Fire extinguisher/equipment available and adequate	Y☐ N☐ n/a☐			
44. Spills containment equipment available	Y☐ N☐ n/a☐			
45. Emergency response procedures available	Y☐ N☐ n/a☐			
46. Emergency training to appropriate personnel	Y☐ N☐ n/a☐			
47. Emergency drills – practised	Y☐ N☐ n/a☐			
48. Emergency meeting point available	Y☐ N☐ n/a☐			
49. First aid kits, available and good condition	Y☐ N☐ n/a☐			
50. First aid officer available	Y☐ N☐ n/a☐			
51. Emergency shower/wash, available and adequate	Y☐ N☐ n/a☐			
Personal protective equipment				
52. PPE being used when required (SWMS)	Y☐ N☐ n/a☐			
53. PPE adequate for exposure	Y☐ N☐ n/a☐			
54. PPE correctly used by employees	Y☐ N☐ n/a☐			
55. PPE in good condition	Y☐ N☐ n/a☐			
56. Protective clothing	Y☐ N☐ n/a☐			
57. Safety shoes, glasses, gloves, hard hats	Y☐ N☐ n/a☐			
58. Fall protection and harness	Y☐ N☐ n/a☐			
59. Eye protection	Y☐ N☐ n/a☐			
60. Face shields	Y☐ N☐ n/a☐			
61. Respirators	Y☐ N☐ n/a☐			
62. Hearing protection	Y☐ N☐ n/a☐			
63. Sun protection	Y☐ N☐ n/a☐			
Hazardous substances				
64. Hazardous substances register current	Y☐ N☐ n/a☐			
65. MSDS available	Y☐ N☐ n/a☐			
66. SWMS lists precautions for use	Y☐ N☐ n/a☐			
67. Substances legibly labelled	Y☐ N☐ n/a☐			
68. Substances properly stored	Y☐ N☐ n/a☐			
Electrical safety				
69. Electrical equipment register	Y☐ N☐ n/a☐			
70. Electrical equipment tested and tagged	Y☐ N☐ n/a☐			
71. Portable generator fitted RCD	Y☐ N☐ n/a☐			
72. Portable residual current device (RCD) tested/tagged	Y☐ N☐ n/a☐			
73. No exposed electrical cords	Y☐ N☐ n/a☐			
74. Power tools do not have cracked or broken insulation, plugs and casings	Y☐ N☐ n/a☐			
75. Plugs, cords and receptacles of power tools are inspected before use	Y☐ N☐ n/a☐			

76. Correct tools used for the job	Y ☐ N ☐ n/a ☐		
77. All operators qualified	Y ☐ N ☐ n/a ☐		
78. Electrical PPE available for 'authorised operators"	Y ☐ N ☐ n/a ☐		
79. Fuses provided	Y ☐ N ☐ n/a ☐		
80. Boundary-off electrical work area	Y ☐ N ☐ n/a ☐		
81. Proper fire extinguishers provided	Y ☐ N ☐ n/a ☐		
82. Electrical danger/high voltage signage displayed	Y ☐ N ☐ n/a ☐		
83. Nearby personnel adequately protected	Y ☐ N ☐ n/a ☐		
Plant and equipment			
84. Plant register current	Y ☐ N ☐ n/a ☐		
85. Regular maintenance/services provided	Y ☐ N ☐ n/a ☐		
86. Daily log books completed	Y ☐ N ☐ n/a ☐		
87. Operators ticketed/competency verified	Y ☐ N ☐ n/a ☐		
88. SWMS followed	Y ☐ N ☐ n/a ☐		
89. Flagmen used where required	Y ☐ N ☐ n/a ☐		
90. Speed limits observed	Y ☐ N ☐ n/a ☐		
91. Weight limits and load sizes observed	Y ☐ N ☐ n/a ☐		
92. Plant parts are safe and damage free	Y ☐ N ☐ n/a ☐		
93. Proper fire extinguishers provided	Y ☐ N ☐ n/a ☐		
94. Plant operating environment/site is safe	Y ☐ N ☐ n/a ☐		
95. Traffic control adequate	Y ☐ N ☐ n/a ☐		
Lifting devices (hoists, cranes, derricks)			
96. Rented cranes inspected and all deficiencies corrected before use	Y ☐ N ☐ n/a ☐		
97. Inspection and maintenance logs maintained	Y ☐ N ☐ n/a ☐		
98. Operators licensed	Y ☐ N ☐ n/a ☐		
99. Crane charts in cab	Y ☐ N ☐ n/a ☐		
100. Equipment firmly supported and outriggers used fully	Y ☐ N ☐ n/a ☐		
101. Powerlines de-energised or clearances maintained	Y ☐ N ☐ n/a ☐		
102. Signalmen work as instructed and trained	Y ☐ N ☐ n/a ☐		
103. Proper signals understood and posted	Y ☐ N ☐ n/a ☐		
104. Swing radius barricades used	Y ☐ N ☐ n/a ☐		
105. High wind operating restriction level applied	Y ☐ N ☐ n/a ☐		
106. Chokes, chains, slings and shackles used correctly	Y ☐ N ☐ n/a ☐		
Storage areas			
107. Accessibility to storage area is proper	Y ☐ N ☐ n/a ☐		
108. Storage shelve in good condition	Y ☐ N ☐ n/a ☐		
109. Lifting & stacking aids in use	Y ☐ N ☐ n/a ☐		
110. Lifting and stacking aids in good condition	Y ☐ N ☐ n/a ☐		
111. Lighting and ventilation in storage area adequate	Y ☐ N ☐ n/a ☐		
112. Warning signs displayed	Y ☐ N ☐ n/a ☐		
113. Hazardous substances are isolated and labelled	Y ☐ N ☐ n/a ☐		
114. Chemicals/flammables stored correctly	Y ☐ N ☐ n/a ☐		
115. Stacked materials in shelves are stable	Y ☐ N ☐ n/a ☐		
116. Waste and trash disposed of regularly	Y ☐ N ☐ n/a ☐		
117. Spills cleaned up promptly	Y ☐ N ☐ n/a ☐		
118. Deluge shower/wash points available	Y ☐ N ☐ n/a ☐		
119. Proper fire extinguishers provided	Y ☐ N ☐ n/a ☐		
120. PPE used in material handling	Y ☐ N ☐ n/a ☐		

continued ...

Exhibit 14 continued

Height works			
121. Perimeter protection for upper floors adequate	Y ☐ N ☐ n/a ☐		
122. Penetrations covered/barricaded	Y ☐ N ☐ n/a ☐		
123. Fall restraint/arrest systems in use	Y ☐ N ☐ n/a ☐		
124. Scaffolds erected on solid footing	Y ☐ N ☐ n/a ☐		
125. Scaffold erection properly supervised	Y ☐ N ☐ n/a ☐		
126. Scaffold structural members are free from defects and meet safety requirements	Y ☐ N ☐ n/a ☐		
127. Scaffolds tied to structure	Y ☐ N ☐ n/a ☐		
128. Guardrails, intermediate rails and toe boards in place	Y ☐ N ☐ n/a ☐		
129. Ladders used are sufficient for the task	Y ☐ N ☐ n/a ☐		
130. Ladders secured to prevent slipping, sliding or falling	Y ☐ N ☐ n/a ☐		
131. Metal ladders are not used near electrical equipment	Y ☐ N ☐ n/a ☐		
132. Working areas free of debris, snow, ice, grease etc.	Y ☐ N ☐ n/a ☐		
133. Workers protected from falling objects	Y ☐ N ☐ n/a ☐		
134. SWMS followed in height works	Y ☐ N ☐ n/a ☐		
Manual handling			
135. Trolleys/lifting aids in use	Y ☐ N ☐ n/a ☐		
136. Workers are trained in proper lifting/carrying techniques	Y ☐ N ☐ n/a ☐		
137. Workers are trained in safe use of lifting aids	Y ☐ N ☐ n/a ☐		
138. Heavy material not carried up or down stairs/ladders	Y ☐ N ☐ n/a ☐		
139. Team lifting is used for large/heavy loads	Y ☐ N ☐ n/a ☐		
140. Material loading and offloading points are kept between knees and shoulder level	Y ☐ N ☐ n/a ☐		
141. SWMS followed	Y ☐ N ☐ n/a ☐		
142. Job rotations undertaken	Y ☐ N ☐ n/a ☐		
143. Awkward postures (leaning forward/stooping) are not required	Y ☐ N ☐ n/a ☐		
144. Bending or twisting is not required to take materials	Y ☐ N ☐ n/a ☐		
Air quality control			
145. Dust suppressed/watered down continuously	Y ☐ N ☐ n/a ☐		
146. Stock piles protected from wind continuously	Y ☐ N ☐ n/a ☐		
147. Plant and equipment maintained to minimise emissions	Y ☐ N ☐ n/a ☐		
148. Worksite is ventilated where volatile organic compounds-based materials used	Y ☐ N ☐ n/a ☐		
149. Perimeter of the site screened to a sufficient height or sources of dust are screened adequately	Y ☐ N ☐ n/a ☐		
Noise control			
150. The quietest plant and equipment used on site	Y ☐ N ☐ n/a ☐		
151. Plant and equipment are maintained in good mechanical order and fitted with silencers/mufflers	Y ☐ N ☐ n/a ☐		
152. Noisy works identified	Y ☐ N ☐ n/a ☐		
153. Hearing protection used (SWMS)	Y ☐ N ☐ n/a ☐		
154. Site hours observed	Y ☐ N ☐ n/a ☐		

Waste management				
155. Waste reduction plan in place	Y☐ N☐ n/a☐			
156. Waste contractor records available	Y☐ N☐ n/a☐			
157. Suitable bins provided on site (for litter/cigarette butts/other)	Y☐ N☐ n/a☐			
158. Hazardous wastes captured and disposed of correctly, e.g. paint sludge/ contaminated soil/other	Y☐ N☐ n/a☐			
159. Waste storage and area fenced properly	Y☐ N☐ n/a☐			
160.Ultimate disposal is done certified personnel	Y☐ N☐ n/a☐			
161.Natural waterways protected from site wastes	Y☐ N☐ n/a☐			
Permit-to-work				
162. Permit-to-work procedures in place	Y☐ N☐ n/a☐			
163. High-risk works display permits-to-work (excavation, roof work, confined space, hot works, energy isolation, asbestos works, restricted area works, etc.)	Y☐ N☐ n/a☐			
164.Records of permits-to-work maintained	Y☐ N☐ n/a☐			
Safety training All employees have:				
165. General industry (safety awareness) training	Y☐ N☐ n/a☐			
166. Site-specific induction training	Y☐ N☐ n/a☐			
167. Work activity (SWMS) training	Y☐ N☐ n/a☐			
Public and adjoining structure protection				
168. Work site secured from public access	Y☐ N☐ n/a☐			
169. Adequate public protection is in place where the worksite is near public spaces (floor level protection)	Y☐ N☐ n/a☐			
170. Adequate overhead protection is provided for the public in all areas	Y☐ N☐ n/a☐			
171. Adequate protection provided for adjoining properties/structures	Y☐ N☐ n/a☐			
172.Adequate warning signage in place	Y☐ N☐ n/a☐			
Others (specify)				
Corrective actions approval				
All corrective actions completer are acceptable and approved.				
Approved by (OHS audit team manager)			Date	

Source: Adapted from WorkSafe BC (2009b), CSA (2008), HSE (2010a).

be reported internally to enable investigation and future incident prevention. There are usually legal requirements governing external reporting and notification of incidents. Dangerous occurrences, for example, excavation collapses and deaths, must immediately be reported to the OHS authority. Incidents that result in workers' compensation claims should be reported to the insurer within the stipulated time frame. Damage to public utility lines needs to be reported to relevant utility companies.

• Recording all injuries with adequate details of victims, nature of injury, treatment provided and rehabilitation requests.

• Conducting a thorough incident investigation immediately after receiving the incident report from the site. A systematic approach to incident investigation should be adopted to analyse what occurred, identify immediately the root causes of the incident, establish appropriate corrective actions and define a plan to prevent recurrences.

- Enrolling injury victims into a rehabilitation programme (also known as Return-to-Work Programme).
- Providing counselling to injured workers, their family members and/or witnesses of workplace incidents.

Hence, maintaining good reporting and record-keeping protocols in incident management is critical for a builder. They can serve two vital purposes: (1) they enable structured analyses on incident data to identify patterns in incidents and evaluations of the effectiveness of prevention measures implemented on site, and (2) they help to demonstrate reasonable and responsible responses by the builder to incidents as stipulated by the OHS legislation. Four templates have been developed to support incident management, as shown in Exhibits 15 to 18, based on a content analysis of various documents on incident management.

Incident report					
Project					
Incident reference					
Date of Incident		Time of Incident		am ☐ pm ☐	
Nature of Incident	☐ Injury ☐ Property/Plant Damage ☐ Near Miss ☐ Environmental ☐ Other...............				
Location of Incident					
Description of Incident					
Details of emergency responses					
Details of injury to people					
Details of damage to equipment/property?					
Witness name			Witness contact		
Person reporting Incident	Name:		Position:		
Incident investigation required	☐ Yes ☐ No				
Notification to authorities required	☐ Yes ☐ No				

Exhibit 15 Incident reporting form
Source: Evolved from Hinze *et al.* (1998), Abdelhamid and Everett (2000)

Injury report		
General details		
Project		
Incident reference		
Person reporting injury	Name:	Position:
Details of injured		
Full name		
Home address		
Date of birth		Male ☐ Female ☐
Occupation		
Employer's name		
Employer's address		
Details of injury		
Date of injury		Time of injury		am ☐ pm ☐
Activity in which the person was engaged at the time of injury		
Exact location where injury occurred		

Nature of injury

☐ Fractures/dislocation	☐ Poisoning and toxic effects of substances	☐ Wounds
☐ Sprains/strains	☐ Confusions/bruising/crushing	☐ Amputations
☐ Concussion	☐ Grazes/abrasions	☐ Electric shock
☐ Lacerations/open cuts	☐ Foreign bodies	☐ Multiple injuries
☐ Damage to glasses, hearing aids, etc.	☐ Burns and scalds	☐ Bites/stings
☐ Occupational overuse/injuries	☐ Puncture/penetration	Others (specify)

Apparent type of illness

☐ Epilepsy	☐ Circulatory system/heart problem	☐ Deafness
☐ Dermatitis/skin rashes	☐ Respiratory system/breathing problems	☐ Headaches
☐ Dizziness/fainting	☐ Infectious and parasitic diseases	☐ Shock
☐ Nausea/vomiting	☐ Psychological disorders (e.g. stress)	Others (specify)

Bodily location of injury

☐ Eye	☐ Arm	☐ Internal organs
☐ Ear	☐ Hand and fingers	☐ Multiple parts
☐ Face	☐ Hip	☐ Feet and toes
☐ Head (other than eye and face)	☐ Trunk	☐ Leg
☐ Shoulder	☐ Back	☐ General/unspecified location
☐ Neck	☐ Psychological	Others (specify)

Agency (What caused the injury?)

☐ Animal(s)	☐ Ground and pathways	☐ Person/People
☐ Chemical(s)	☐ Hand tools (non powered)	☐ Slips/trips/falls
☐ Physical environment	☐ Road transport (cars, bikes etc)	☐ Inadequate training
☐ Insect(s)	☐ Equipment (includes powered tools)	☐ Foreign bodies
☐ Floors and passageways	☐ Syringes	☐ Objects
☐ Stairs	☐ Biological agencies	Others (specify)
☐ Fixed or mobile plant/machinery	☐ Manual handling	

Details of treatment			
Treatment provided by first aid officer	Yes☐ No☐	Remarks:	
Name of first aid officer			
Follow up treatment required	Yes☐ No☐	If yes, an incident investigation report must be completed with 24 hours. Notify incident investigator automatically	
Doctor/medical centre attended			
Date attended		Medical Certificate Received	Yes ☐ No ☐
Treatment (i.e. x-ray, prescription)			
Further consultation required	Yes☐ No ☐	Rehabilitation programme required	Yes ☐ No ☐
Days off from work			
Details of permanent incapacity(if any)			

Exhibit 16 Injury reporting form
Source: Modified from Toole (2002), CSA (2008)

Incident investigation report					
Details of incident					
Project					
Incident reference					
Incident investigator	Name:			Position	
Date of investigation/report					
Risk rating of incident					

Using the two variable risk matrix at right:

- Rate the severity of the incident
- Rate the likelihood of the incident occurring or recurring
- Circle the resultant risk rating on the risk matrix

Likelihood label	Severity label				
	Negligible	*Significant*	*Moderate*	*Major*	*Catastrophe*
Almost certain	Medium	High	High	Very high	Very high
Likely	Medium	Medium	High	High	High
Possible	Low	Medium	High	High	High
Unlikely	Low	Low	Medium	Medium	High
Rare	Low	Low	Medium	Medium	High

Causes of incident

Behavioural causes
What are the behaviours that contributed to the cause of the incident?

1. Performing task without authority □
2. Performing task at unsafe speed □
3. Performing task while affected by drugs/alcohol □
4. Performing task with improper work technique □
5. Performing task without personal protective equip □
6. Performing task without correct personal protective equip □
7. Failure to warn of hazard □
8. Failure to secure hazardous item □
9. Making safety device inoperable □
10. Distracting, teasing or abusing a person □
11. Using unsafe or tagged out equipment □
12. Using equipment in an unsafe manner □
13. Unsafe placement of equipment or objects □
14. Unsafe manual handling technique □
15. Unsafe position or posture □
16. Unsafe acts of others □
17. Other (specify)
18. Not applicable □

Describe

Management system causes
What are the management systems deficiencies that contributed to the cause of the incident?

19. Inadequate standard operating procedures/policies □
20. Inadequate supervision □
21. Inadequate hazard identification □
22. Inadequate assessment of risk □
23. Inadequate provision of personal protective equip □
24. Inadequate operator training □
25. Inadequate supervisor training □
26. Inadequate fire or explosion risk control □
27. Inadequate noise control □
28. Inadequate ventilation □
29. Inadequate temperature control □
30. Inadequate fall protection □
31. Inadequate signage or warning systems □
32. Inadequately controlled use of chemicals/substances □
33. Other (specify)
34. Not applicable □

Describe

Physical causes What are the physical conditions that contributed to the cause of the incident?	35. Inadequate or absent guarding ☐ 36. Poor workstation design or layout ☐ 37. Poor condition of equipment or objects ☐ 38. Equipment or objects with unsafe design ☐ 39. Unsafe storage of equipment/objects (housekeeping) ☐ 40. Unsafe walking surfaces ☐ 41. Unsafe lighting or glare ☐ 42. Unsafe clothing or shoes ☐ 43. Unsafe task or process ☐ 44. Inadequate fire or explosion risk control ☐ 45. Inadequate noise control ☐ 46. Inadequate ventilation ☐ 47. Inadequate temperature control ☐ 48. Inadequate fall protection ☐ 49. Inadequate signage or warning systems ☐ 50. Inadequately controlled use of chemicals/substances ☐ 51. Other (specify) 52. Not applicable ☐
	Describe
Environmental causes What are the environmental conditions that contributed to the cause of the incident?	53. Unexpected ground conditions ☐ 54. Adverse weather ☐ 55. Natural catastrophe ☐ 56. Others (specify) 57. Not applicable ☐
	Describe
Recommended actions to prevent recurrence	
☐ Elimination ☐ Substitution ☐ Engineering control ☐ Administrative control ☐ Personal protective equipment	
Detail the actions	

Source: Adapted from CSA (2008) and WorkSafe BC (2010).

Exhibit 17 Incident investigation form
Source: Adapted from CSA (2008) and WorkSafe BC (2010)

System development

The newly introduced Google App Engine technology was utilised to develop the proposed OHS monitoring and control system. Google App Engine is a platform as a service (PaaS) cloud computing platform for developing and hosting web applications in the Google-managed data centre, free of cost for small- to medium-scale web applications. Conventionally, data-driven web applications have been developed around a three-tier architecture: client tier–web server tier–database server tier. In contrast, Google App Engine provides a single easier platform for building and hosting web applications without the need for complex web server and database server programming. Google App Engine includes the following features within it to enable single-point web development and hosting:

- dynamic web serving, with full support for common web technologies;
- datastore – persistent data storage with queries, sorting and transactions;
- automatic scaling and load balancing;

Rehabilitation programme					
General details					
Victim's name					
Injury details					
Medical advice provided					
Rehabilitation coordinator	Name			Contact details	
Language interpreter	Name			Contact details	
Rehabilitation programme					
Rehabilitation provider	Name		Organisation		Contact details
Duration of rehabilitation	Start date			End date	
Rehabilitation description					
Rehabilitation consultations	*Date*	*Mode of discussion*	*Matters discussed*	*Next appointment*	
Attachments	*Date*	*Description of attachment*		*Attachment upload*	
Modification of duties					
Pre-injury duties					
Pre-injury work location					
Pre-injury work hours					
Modifications	*Period*	*Modified duties*	*Modified work location*	*Modified work hours*	*Remarks*
	From... To...				

Exhibit 18 Rehabilitation form
Source: Modified from Koehn and Datta (2003) and SafetyNet (2010)

- runtime environment – to control data query, sorting and transactions (a choice of Go, Java or Python runtime environment is available for developers);
- web-based administration console (APIs) – for authenticating users and sending emails using Google accounts; and
- Google web toolkit (GWT) – allows developers to code web application on their computers by emulating all of the App Engine services on a developer's local computer.

In essence, with Google App Engine, there are no servers to maintain; web application developers can just upload their applications, and they will be ready to serve their users.

Google App Engine adopts the Model-View-Controller system architecture concept and Figure 4.2 depicts the model of the proposed system based on this concept.

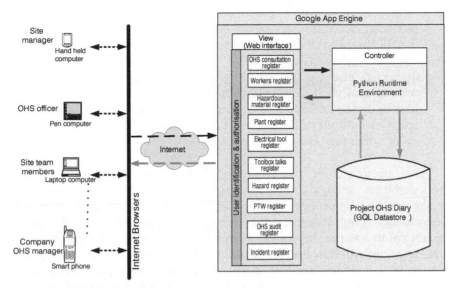

Figure 4.2 System architecture

The web system architecture consists of the following four components:

1   Clients – refers to system users, client computers and web browser software that interact with the system for data entry and information retrieval.
2   Project OHS diary – sits in the Google App Engine datastore, which uses a distributed architecture to automatically manage scaling to a very large dataset. It organises and manages OHS data that is created, and modified by system users.
3   Python Runtime Environment – which controls data query, sorting and transactions between clients and the datastore.
4   View/web interface – which is a set of JavaScript-based forms that facilitates communication between clients and the datastore through the controller.

Key considerations that drove the selection of Google App Engine with Python for the development of the proposed web application are:

1   Cost – Python and App Engine Python API are free to download for implementing web-based systems. Systems can be tested on own machines with minimal set-up before deployed on the web.
2   Flexibility – One of the key advantages of using datastore is that it is designed to scale, allowing applications to maintain high performance as they receive more traffic.

3   Interoperability – App Engine and Python are available to install on any platform. Hence this service can be installed on any computers/servers then deployed to cloud service when needed.

4   Easiness – Python is powerful, very widely used on web service designing. Also Google App Engine Python/NDB API provides easy ways to implement data transactions.

5   Ease of update and bug fixing – Both Python and NDB API are current and have an ongoing development and update policy which is based on needs rather than commercial pressure.

The implementation of the proposed system on Google App Engine was performed using a Google Web Toolkit (GWT). The use of GWT can make web development projects quicker and more productive; it can greatly increase development speed, reduce debugging and testing time, and improve quality of the output. It provides both visual and code-level capabilities for creating standards-based websites and designs without deep knowledge and expertise in web development languages. Consequently, the translation of the system architecture above into a physical web application involved three tasks; the first part dealt with defining the entity kinds of the datastore for the system, the second part concentrated on the process issues dealing with data transactions – the communication flow and workflow management issues within OHS management – and the final part focused around the user interface. The next sections describe these in detail.

Building the datastore

The entire web application and its use are driven by the project OHS diary that sits in the Google App Engine Datastore. Careful and error-free design and implementation of the OHS diary is therefore paramount for the smooth operation of the proposed system. Unlike SQL databases in conventional database servers, the datastore is not a relational database; it is a public API for accessing Google's Bigtable high-performance distributed database system.

A good datastore design starts with the design of a good data model. A data model is a conceptual representation of the data structures that are required by a datastore. The data structures include the data objects, the associations between data objects, and the rules which govern operations on the data objects. As the name implies, the data model focuses on what data are required and how they should be organised rather than what operations will be performed on the data. To use a common analogy, the data model is equivalent to a building plan. A data model is a plan for building a datastore that helps to make sure that all data objects required by the datastore are completely and accurately represented. The data management concepts for the datastore are the same as in other data modelling systems, though the terminology is a bit different. Table 4.1 compares the terminologies used in object-oriented databases, relational databases and datastore; the

Table 4.1 Data modelling terms

Object oriented database	Relational database	Datastore
Class	Table	Kind
Object	Record	Entity
Attribute	Column	Property

Source: Ciurana (2009)

Table 4.2 Datastore hierarchy

Parent Kind	Children Kind		
	Level 1	Level 2	Level 3
Employee			
Project	Incident	Injury	Rehab
	OHS Meeting		
	OHS Audit	OHSAuditAction	
	Plant	Plant Inspection	
		Plant Pre-start Check	
	PTW	PTW_Approval	PTW_Cancel
	Hazard		
	Hazard_Substance		
	Electrical_Equipment		
	Worker	PPE	
	OHSTraining		
	ToolBoxTalks	ToolBoxAction	
		ToolBoxAttendee	
		ToolBoxTopic	
	Client		
	Contractor		

datastore organises data in kinds, which consist of entities that are defined by properties.

A datastore hierarchy for kinds for the proposed web application was developed as shown in Table 4.2. Subsequently, twenty-six kinds were created for the proposed system. Each entity in the datastore has a key that uniquely identifies it and consists of the kind, identifier and ancestor path (for datastore hierarchy).

Building the Python runtime environment

Having built the project OHS diary on datastore, the next step was to define App Engine transactions for manipulating data in the datastore. A transaction is a set of datastore operations on one or more entities. In a web application, the actions to be performed in the transaction are defined using Python functions. Accordingly, Python-based transactions had to be developed for two distinct purposes in the proposed web system, including report generation and workflow management, as described below.

Report generation in the system

Data in the project OHS diary can be processed by predefined transactions to produce desired information to enable effective monitoring and control of OHS performance on site. Cheung *et al.* (2004) suggested a list of OHS indicators that characterise OHS performance in a project, including:

1 OHS compliance – reports/evidence of OHS non-conformance in managing safe work practices, plant and machinery, electrical tools, hazardous substances and personal protective equipment for workers.
2 OHS training – summaries/reports of OHS consultation meetings and tool box talks, and lists of refresher training requirements for project team members.
3 OHS audit outcomes – summaries/reports of number of internal and external OHS audits and their outcomes.
4 Incident statistics – number of incidents, injuries, site closures and lost man days.

As such, transactions were designed to obtain the following sixteen types of critical outputs from the data in the project OHS diary that satisfy the above four categories:

1 A list of plant whose re-registration is falling soon
2 A list of plant whose maintenance is falling soon
3 Service/hazard response request report/notice for plant that failed in regular plant inspection
4 A list of electrical tools whose retests are falling soon
5 Hazardous substances that do not have risk assessment completed
6 Permit-to-work approvals for printing and display
7 Records of activities whose permits-to-work have been cancelled
8 Records of hazard reports received and responses to them
9 A list of project team members whose refresher training is due soon
10 OHS consultation meeting minutes
11 A list of OHS audit items that need the project team's attention
12 Project incident summary
13 Individual incident report
14 Personal injury report
15 Personal rehabilitation report
16 Incident investigation and analysis summary

Workflow management in the system

Apart from effective data capturing and processing, workflow/approval management is another critical aspect involved in OHS implementation on site. It was therefore necessary to identify tasks that involve workflow management and approval processes and design protocols and communication flows in the proposed system. Accordingly, seven tasks in OHS management were identified that involve workflow/approval processes. Communication flow diagrams were developed for these tasks and utilised when programming Python-based transactions for the proposed system. The seven OHS tasks and their respective communications flows are described below.

1   OHS consultation meetings communication flow: an OHS consultation meeting is kicked off by an invitation sent to potential attendees by the meeting organiser, with a meeting agenda. Once the meeting has taken place, the minutes of the meeting are then circulated to the attendees. Any items arising as action items in the meeting are communicated to protect team members for their actions. When these items are attended to, the project team members may notify the OHS meeting organiser of the OHS action taken and they subsequently can update OHS consultation meeting attendees.

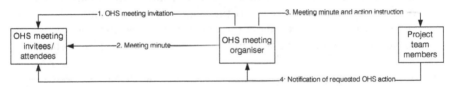

2   Hazard reporting communication flow: hazard reporting can be initiated by anyone who spots a hazard on site (both registered and unregistered users). Once reported, the hazard report is sent to the site manager who then delegates the hazard response task to a site team member. When the hazard has been managed/treated by the site team member, the site manager is notified of the action. The site manager may update the initiator of the hazard report of the action taken. The project OHS officer automatically receives a copy of communications along the entire process.

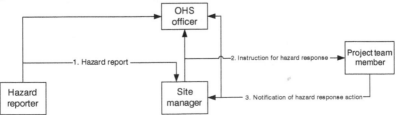

3   Hazardous substances communication flow: material risk assessment request is initiated by the site store keeper; once initiated, the request is sent to the site manager who then delegates the risk assessment task to a project team member. When the assessment has been completed by the project team member, the site manager is notified of it. The project OHS officer automatically receives copies of communications along the entire process.

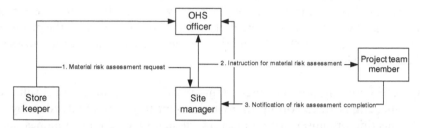

4   Regular plant inspection communication flow: regular plant inspections are carried out to identify any possible hazards associated with plant and machinery used in a project. At the conclusion of a regular plant inspection, a report is sent by the plant inspector to the site manager who analyses the report and identifies any actions required to respond to the hazards. Subsequently, the project team members are instructed by the site manager to react to the plant inspection report. When the project team members have attended to the issues, they notify the site manager of their actions who may then notify the plant inspector.

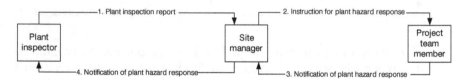

5   Permit-to-work request communication flow: when a task requires a permit-to-work before commencing on site, the relevant project team member initiates a PTW request from the nominated PTW approver/ OHS officer who assesses risk control measures implemented for the task and then either approves or disapproves the PTW request. The approval or disapproval is notified to the PTW requester.

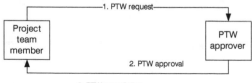

A task that originally obtained a PTW approval may attract a cancellation of PTW during the course of construction if the PTW approver's latest assessment finds any inadequacies in risk management. In such circumstances, the PTW approver may cancel the PTW and this is also notified to the original PTW requester.

6   OHS audit communication flow: OHS audits are carried out to assess the effectiveness of the OHS management system on site and its communication flow has a similar pattern to the plant inspection communication flow. At the end of an OHS audit, a report is sent by the OHS audit team to the site manager who analyses the report and identifies any actions required to rectify the flaws in the OHS management system. Subsequently, project team members are instructed by the site manager to react to the OHS audit report. When the project team members have attended to the issues, they notify the site manager of their responses and these may subsequently be notified to the OHS audit team.

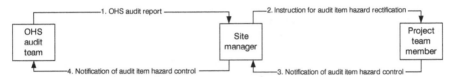

7   Incident management communication flow: when an incident happens on site, an incident report is sent to the OHS officer who assesses the situation and decides if notifications to external organisations are required under the law and then acts accordingly. Alongside this, the incident report may be forwarded to an incident investigator by the site manager, requesting a detailed investigation of the incident. The incident investigator submits a detailed report upon completion of the investigation to the site manager.

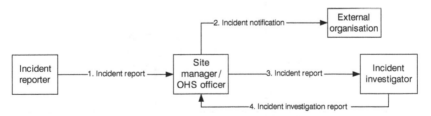

However, the external notification can be done outside the proposed system, using the protocols and proforma suggested by the relevant organisations.

Building the view (web interface)

The interface is critical to the successful adoption of any web application. In a data-driven web application like the proposed system, it is the medium that enables effective data capture, communication and report presentation. Thus, extra care was taken to ensure usage-centred and user-centred interfaces were designed for the system. The user interface for the proposed system was designed to consist of ten pages to deal with different OHS matters in a project, together with the pages for user log-in, project dashboard and personal to-do-list. Figure 4.3 illustrates the arrangement of these pages in the system and the user navigational flow. As shown in the figure, upon a user keying-in the correct username and password, the user is directed to a to-do-list page that shows tasks awaiting the user's actions. The user can then navigate to appropriate pages to complete the tasks. The user may also initiate a new task on any of the pages or view OHS reports.

The creation of the user interface was facilitated by the Google Web Toolkit, which offers a GWT user-interface library. The GWT library consists of widgets and panels to enable easy implementation of user interfaces. Figure 4.4 shows screenshots of some user interfaces in the proposed web application.

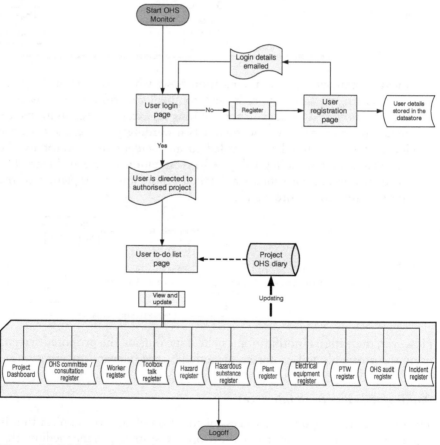

Figure 4.3 Web interface layout

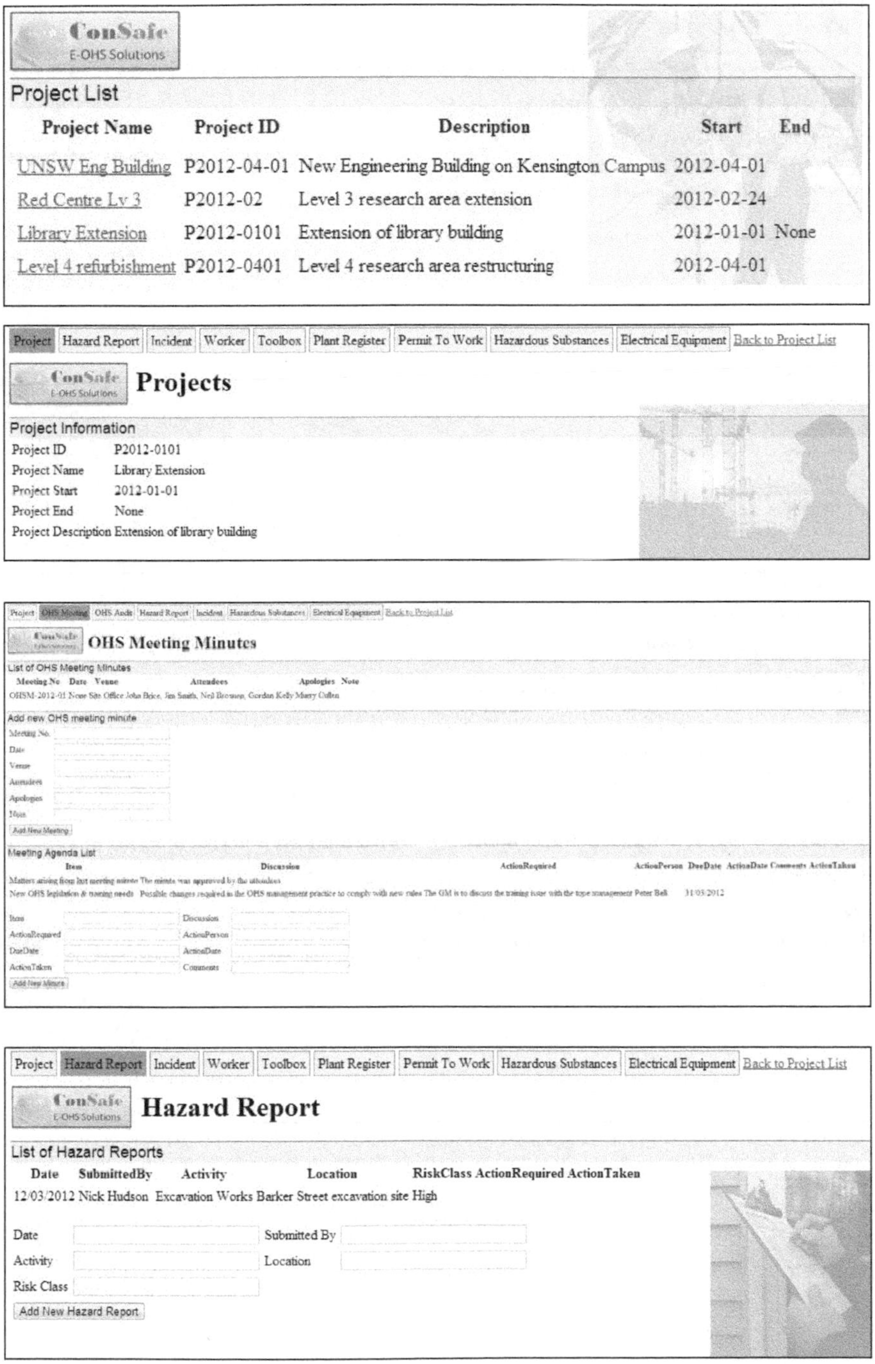

ConSafe
E-OHS Solutions

Project List

Project Name	Project ID	Description	Start	End
UNSW Eng Building	P2012-04-01	New Engineering Building on Kensington Campus	2012-04-01	
Red Centre Lv 3	P2012-02	Level 3 research area extension	2012-02-24	
Library Extension	P2012-0101	Extension of library building	2012-01-01	None
Level 4 refurbishment	P2012-0401	Level 4 research area restructuring	2012-04-01	

Project | Hazard Report | Incident | Worker | Toolbox | Plant Register | Permit To Work | Hazardous Substances | Electrical Equipment | Back to Project List

ConSafe E-OHS Solutions **Projects**

Project Information

Project ID — P2012-0101
Project Name — Library Extension
Project Start — 2012-01-01
Project End — None
Project Description Extension of library building

Project | OHS Meeting | OHS Audit | Hazard Report | Incident | Hazardous Substances | Electrical Equipment | Back to Project List

ConSafe E-OHS Solutions **OHS Meeting Minutes**

List of OHS Meeting Minutes

Meeting No	Date	Venue	Attendees	Apologies	Note
OHSM-2012-01		New Site Office	John Price, Jim Smith, Neil Browne, Gordon Kelly Mary Cullen		

Add new OHS meeting minute
Meeting No.
Date
Venue
Attendees
Apologies
Note
Add New Meeting

Meeting Agenda List

Item	Discussion	ActionRequired	ActionPerson	DueDate	ActionDate	Comments	ActionTaken
Matters arising from last meeting minute	The minute was approved by the attendees						
New OHS legislation & training needs	Possible changes required in the OHS management practice to comply with new rules The GM is to discuss the training issue with the top management		Peter Bell	31/03/2012			

Item
ActionRequired
DueDate
ActionTaken
Add New Minute

Discussion
ActionPerson
ActionDate
Comments

Project | Hazard Report | Incident | Worker | Toolbox | Plant Register | Permit To Work | Hazardous Substances | Electrical Equipment | Back to Project List

ConSafe E-OHS Solutions **Hazard Report**

List of Hazard Reports

Date	SubmittedBy	Activity	Location	RiskClass	ActionRequired	ActionTaken
12/03/2012	Nick Hudson	Excavation Works	Barker Street excavation site	High		

Date
Activity
Risk Class
Add New Hazard Report

Submitted By
Location

Figure 4.4 Web interface examples

continued …

Figure 4.4 continued

| Project | Hazard Report | Incident | Worker | Toolbox | Plant Register | Permit To Work | Hazardous Substances | Electrical Equipment | Back to Project List |

ConSafe E-OHS Solutions **Worker & PPE Record**

Worker List

Name	Address	Mobile	Employer
Sean Watson	39 Balmain Road, Homebush	0421167254	ABC Constructions

List of PPE Checks

Date	Checked By	Alcohol/Drug Test	Competence Test	Work Activity	Comments	Hard Hat	Safety Footwear
12/03/2012	Yes	Yes					Yes

Add new PPE Check

Date

Checked By

Alcoho/Drug Test

Competence Test

Work Activity

Comments

Hard Hat

Safety Footwear

Add PPE check record

| Project | Hazard Report | Incident | Worker | Toolbox | Plant Register | Permit To Work | Hazardous Substances | Electrical Equipment | Back to Project List |

ConSafe E-OHS Solutions **Plant Register**

List of Plant Registry

Plant Name	Serial Number	Make	Plant Provider	Registry Expiry	Service Date	Service Frequency	Service Due	Log Book Available
Excavator	Plant001	Catapillar330	Owned Plant	10/10/2012	01/08/2012	Biannually	01/04/2012	Yes

PreStart Check

Date	Operator	Inspector	Visibility	Cabin Safety	Brake Condition	Remarks
12/03/2012	Jim Baker	Gordan Kelly	Yes	Yes	No	

Date   Operator
Inspector   Visibility
Cabin Safety   Brake Condition
Remarks
Submit

Plant Inspection

Inspector	Date	Risk Assessment	Operation Manual	Maintenance Report	Remarks	Manager	Action Required	Action Due Date	Supervisor	Action Taken	Action Date
Neil Brosnen	10/03/2012	Yes	Yes	No		Gordan Kelly	Find the maintenance report	12/03/2012	Jim Smith		

Inspection
Date   Inspector
Risk Assessment   Operation Manual
Maintenance Report   Manager
Action Required   Action Due Date
Supervisor   ActionTaken
Submit

Project | Hazard Report | Incident | Worker | Toolbox | Plant Register | Permit To Work | Hazardous Substances | Electrical Equipment | Back to Project List

ConSafe
E-OHS Solutions
Permit To Work Register

List of PTW
Location	ValidFrom	ValidUntil	Task
Basement 1-east side	12/03/2012	12/04/2012	

Location [   ]   Valid From [   ]

Valid Until [   ]

[Add PTW Request]

List of PTW Approval
Date	ApprovedBy
11/03/2012	Neil Brosnen

List of PTW Cancellation
Date	CancelledBy

Permit to work
All contractors must obtain
a valid permit to work on
this site

Project | OHS Meeting | OHS Audit | Hazard Report | Incident | Hazardous Substances | Electrical Equipment | Back to Project List

ConSafe
E-OHS Solutions
OHS Audit

List of OHS Audits
AuditID	Date	AuditTeam	ReportSentTo	General	HouseKeeping	PPE	Comment
2012-A03	20/03/2012	undefined	John Brice	OHSE policy has not been updated	Unused ladder has been left on the scaffolding on the basement floor	N/A	undefined

Audit ID [   ]   Date [   ]

AuditTeam [   ]   Report Sent To [   ]

General [   ]   HouseKeeping [   ]

PPE [   ]   Comment [   ]

[Add OHS Audit]

OHS Audit Action Items
Action	ApprovedBy	Date
Policy updated, ladder removed	John Paul	31/03/2012

Action [   ]   ApprovedBy [   ]

Date [   ]

[Add Action Item]

Project | Hazard Report | Incident | Worker | Toolbox | Plant Register | Permit To Work | Hazardous Substances | Electrical Equipment | Back to Project List

ConSafe
E-OHS Solutions
Incident Register

List of Incidents
Reference ID	Date	Time	Nature	Location	Description	Emergency Response	Injured Person	Damaged Equipment	Witness Name	Witness Contract	Reported By	Investigation Required	Notification Required
IR01	19-03-2012	2pm	Fall from scaffold	Basement 1-east side	A worker fell off a scaffold while applying waterproofing	First aid, called ambulance	Sean Watson	Scaffold braces	Alan Bold	0421687735	Neil Brosnen	Yes	Yes

Reference ID [   ]   Date [   ]

Time [   ]   Nature [   ]

Location [   ]   Description [   ]

Emergency Response [   ]   Injured Person [   ]

Damaged Equipment [   ]   Witness Name [   ]

Witness Contact [   ]   Reported By [   ]

Investigation Required [   ]   Notification Required [   ]

[Report new incident]

Injury List
Name	Date	Activity	Location	Nature	Treatment

Name [   ]   Date [   ]

Activity [   ]   Location [   ]

Nature [   ]   Treatment [   ]

[Submit]

Rehabilitation List
Provider	StartDate	EndDate	Plan	PreDuties	Modifications

Provider [   ]   StartDate [   ]

EndDate [   ]   Plan [   ]

PreDuties [   ]   Modifications [   ]

[Submit]

Evaluation of the system

An evaluation exercise was carried out to assess the effectiveness of the proposed web-based OHS monitoring system in achieving the desired goals as outlined in the introduction section of this chapter. In this context, it aimed to assess the following.

1   The adequacy of functionalities in the proposed system prototype for managing essential components of OHS monitoring and tracking.
2   The level of support provided by the proposed system for collaborative OHS monitoring and tracking, OHS compliance and continual OHS improvement.
3   Challenges that could be overcome by the proposed system, which are inherent in the traditional approach to OHS monitoring.

Hennink *et al.* (2011) suggested that focus group discussions are a good form of qualitative research method that can be used for exploratory, explanatory or evaluative research. They further highlighted two important elements of successful focus groups – homogeneity among participants and group size (maximum of eight members). A focus group approach was adopted for the evaluation of the proposed system and involved four experienced professionals from the construction industry. For confidentiality reasons, their details are not disclosed here. The system was demonstrated to the participants and their views of the functionalities and performance of the system as well as its possible implications for the OHS management process were sought. The following summarises the comments made by them in the discussion that lasted for about 80 minutes.

1   The proposed system can assist with real-time capture of OHS implementation data from sites and tracking progress on critical health and safety actions.
2   The system can remove the complexities in monitoring multiple tasks done by various subcontractors on site.
3   It can remove the burden of recording OHS details on paper and safekeeping them.
4   With effective record-keeping practices facilitated by the system, builders can easily demonstrate their OHS compliance and avoid penalties and prosecutions by OHS authorities.
5   Organised OHS data capturing, processing and communicating fostered by the proposed system could enable OHS performance measurement and continual improvement.
6   Organisational learning of OHS from mistakes and incidents can be facilitated with the availability of OHS performance information provided by the system.

7   Organisational learning often leads to innovation. Thus, learning driven by an effective monitoring and tracking system like the proposed system could enable innovation in OHS practices.

It can be summarised that the proposed system satisfies the requirement for a comprehensive online diary for OHS monitoring and control on construction sites.

Conclusion

The successful monitoring and tracking of occupational health and safety (OHS) management systems on construction sites is a critical step in the prevention of accidents. There are many challenges to effective deployment of OHS monitoring and tracking systems, including: constantly changing hazardous conditions as a project progresses, multilayered subcontracting and losing track of OHS responsibilities, concurrent progresses of multiple tasks on site, abundance of hazardous material use, plant operations and tradesmen, limitations of supervisors to monitor every aspect meticulously, and a large amount of paperwork and risk of losing records. However, capabilities that ICT provides can assist builders in overcoming these challenges and improving OHS monitoring process. Although collaborative project management using project extranets/web portals has reached a reasonably mature state in construction, its uptake in OHS management is still at a primitive stage. A groupware system was developed to enable collaborative OHS monitoring in construction through a platform web. The new groupware system could deliver numerous benefits to builders. (1) Real-time capture of OHS implementation data from sites helps track actions and measure progress on critical health and safety actions. (2) With effective record-keeping practices, builders can easily demonstrate their OHS compliance and avoid penalties and prosecutions by OHS authorities. (3) Organised OHS data capturing, processing and communicating enable OHS performance measurement and continual improvement. (4) Organisational learning of OHS from mistakes and incidents is fostered, leading to innovation in OHS practices.

5 Virtual community of safety practice for construction

Introduction

Informal conversations occur all day long among employees in an organisation – in the lunchroom, on a coffee break or in corridors. These conversations allow employees to share experiences. The shared experience may be an innovative idea, procedure or an insight into how a person performs a task more effectively and efficiently. These experiences embody expertise and know-how and are known as tacit knowledge, which becomes an asset for the organisation. Hence, it is crucial for an organisation to create a structure to enable internalisation of this valuable asset. One way to create a structure among employees is to develop a community of practice (Wenger *et al.* 2002). A community of practice is a group of people who share an interest or passion for something that they know how to do, and who interact regularly in order to learn how to do it better. The primary purpose of communities of practice is to create and share knowledge among participants. Carlsson (2003) quoted that communities of practice enhance and improve effectiveness of both individuals and the organisation.

Communities of practice could offer a great deal of opportunity and potential for improving the occupational health and safety performance of builders. Lingard and Rowlinson (2005) claimed that a construction company may have several professionals and team players. Each professional may have some knowledge and experience in OHS. If these experiences and knowledge were collated and internalised, it could help improve organisational learning ability and thereby OHS performance. Likewise, Chua and Goh (2004) argued that, in order for the construction industry to improve its poor safety performance, it needs to learn from its mistakes and put the lessons learned to good use. Gherardi and Nicolini (2000) reinforced that safety is situated practice and safety knowledge is culturally mediated by forms of social participations, material working conditions and the negotiated interpretations of actions on site. Safety knowledge is therefore dynamic and profoundly rooted in communities of practice. Safety culture is learned when joining a community of practice as a distinctive feature of professional identity. Hence, the formation of communities of safety practice, amalgamating professionals from different projects, could facilitate the sharing of tacit OHS knowledge and thereby improve safety performance of construction organisations and industry as a whole. Moreover, this could assist the organisation in alleviating the challenges posed by high employee turnover and skill shortages. It is also an effective means of educating new entrants to the community or organisation. The community could serve as a viable source of information on new and proposed regulations, successes related to best practices, new concerns of interest to safety professionals, accident reviews, emerging issues or concerns, etc. Another means of getting information would be through enquiries made by individuals to the entire community of safety practice. Flannery and Hinze (2008) also underlined that through participation in a community of safety practice, which consists of safety professionals who network regularly through meetings and conferences, this type of information could be shared on an ongoing basis.

The application of communities of practice in construction faces a major challenge. Because construction projects are remote and scattered, and each project has its set completion schedule and progress status, interactions between employees of construction organisations are marginal. Hence, it is impractical to establish a group of safety professionals from different construction sites who could meet on a regular basis to share information of mutual interests and to learn from past mistakes. Nonetheless, the formation of online/virtual communities of practice, departing from the conventional model, leveraging on the power of Web 2.0 technologies, would alleviate the impediments caused by geographical, time and work pressure constraints. Moreover, the information shared by community members could systematically be archived for future reference, to avoid reinventing the wheel.

This chapter reports on the development of a virtual community of safety practice (VCoSP) for construction organisations, using Web 2.0 technologies, to provide a platform for realising the benefits indicated above and thereby improving safety performance in the construction industry at large. The development process of the VCoSP traversed through the following stages:

1   Establishing the terms of reference for the virtual community of safety practice that describe the nature of knowledge content, mode of knowledge creation and dissemination, and the components and features needed in the system to achieve the ultimate goal.
2   Developing the virtual system for the established terms of reference using a Web 2.0 tool.
3   Validating the system with construction safety practitioners to assess its effectiveness in assisting to improve safety in construction.

The research process and findings are discussed under different sections in the chapter. First, literature reviews on communities of practice and virtual communities of practice are presented, and then the terms of reference for the proposed VCoSP. Next the development and validation of the system is discussed, followed by conclusions.

Communities of practice

Many definitions are found in the literature for communities of practice. Each definition tells about varying features of communities of practice. Wenger *et al.* (2002) defined that communities of practice (CoPs) are groups of people that share a concern, a set of problems or a passion about a topic and who deepen their knowledge and expertise in this area by interacting on an ongoing basis. Likewise, McDermott (1999) stated that CoPs are groups of people who share ideas and insights, help each other solve problems and develop a common practice or approach to the field. These groups are formed to share what they know and to learn from one another regarding some aspects of their work (Nickols 2003). They formalise their existence through the establishment of common goals and values and are often deliberate in their construction and seek to meet predetermined needs which have been identified by their participants (Molphy *et al.* 2007). Hubert *et al.* (2001) and Kimble and Barlow (2000) held that CoPs are learning environments. They are groups of people who come together to share with and learn from one another. They are held together by a common interest in a body of knowledge, and are driven by a desire and need to share problems, experiences, insights, templates, tools and best practices. CoPs are an intrinsic condition for the existence of knowledge. They are a tool for converting implicit knowledge into an explicit form of knowledge (Davenport and Prusak 1998). Knowledge-intensive consulting firms value CoPs as a valid method for knowledge acquisition and transfer, as key elements in CoPs are knowledge sharing and learning. The learning that evolves from these communities is collaborative, and the collaborative knowledge of the community is greater than any individual knowledge. In a CoP environment, a person's learning is enhanced through engagements with others which enable the extension of that person's capability to a new, high level.

Wenger (1998) defined the life cycle of CoPs as having five maturity stages, as shown in Figure 5.1.

1   Potential stage – involves finding people with similar interests, establishing contacts and building informal relations.
2   Coalescing stage – where identity is formed and the values are discussed. Members engage in discussions in the field of interest and move from a loose network to a common sense of purpose.
3   Active stage – where a CoP becomes highly dynamic and comes into its own by engaging in high-level interactions and socialisation. This is where generation of new knowledge, dissemination and learning occurs.

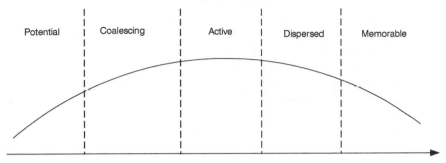

Figure 5.1 Life cycle of CoPs

4   Dispersed stage – members of the community lose interest in the topic, activities reduce and influx of new knowledge is lessened, making the CoP less attractive.
5   Memorable stage – here the CoP is dispersed but tales and anecdotes live on for a while. People still associate with the CoP as part of their identity.

Virtual communities of practice

A virtual community of practice (VCoP) is a network of individuals who share a domain of interest about which they communicate online. The practitioners share experiences, problems and solutions, tools and methodologies, online, contributing to the development of the knowledge of each participant in the community as well as the domain as a whole. VCoPs enhance the learning environment since they allow both synchronous and asynchronous communications, integrate geographically isolated experts with novice, and promote situated learning. Here is a summary of several benefits of VCoPs, as reported in the literature.

- Physical location of employees is unimportant and isolation from the peer group is less problematic when employees are scattered geographically. The use of technology bridges the geographical gaps. Hence, virtual CoPs enable collaborations, sharing of specialist interests and experience, and access to mentors and like-minded individuals, irrespective of geographical location.
- Virtual CoPs can afford a combination of synchronous and asynchronous communications and socialisation, access to and from geographically isolated communities, and international information sharing.
- Virtual CoPs and ICTs provide a systematic means for managing knowledge as an asset. Experiences, insights and ideas of people are captured and stored systematically, which can be reused in the future via sound search and retrieval functions.
- Virtual CoPs provide interested professionals with opportunities for collaboration, discussions and debates through networked technologies.

This gives them the benefit of others' experiences and often saves them from reinventing the wheel as they can find out what others have done when faced with particular problems.

- Virtual CoPs establish a networked environment where interactions that improve learning can occur. The interactions within these communities focus around knowledge sharing within the membership, who may range from experts to novices. Via the interactions of experts and novices, neo-apprentice-style learning can occur.
- Using ICT to support the ongoing interactions and activities of CoP members frees the members from constraints of time and space.

Critical success factors

Gannon-Leary and Fontainha (2007) identified the following nine critical success factors for the existence and evolvement of virtual CoPs and thereby knowledge creation and learning.

1   Virtual CoPs are formed on ICT and Web 2.0 platforms. The success and growth of VCoPs is heavily reliant on the level of technological provision and ICT skills that members possess to support mutual engagement.
2   The evolution of a CoP is reliant upon the effective communication of the members, most easily achieved through face-to-face meetings. Virtual communities transform personal interactions and physical relationships into cyber interactions and electronic relationships. Hence, technology needs to be regarded as an acceptable and transparent means of communication.
3   Membership is consolidated and trust developed through effective personal identification modes in the virtual CoP.
4   Virtual CoP members must have a sense of belonging – being an insider of the community and actively participating in the activities.
5   A virtual CoP must have a purpose and this purpose must be achievable via ICTs.
6   Consideration needs to be given to the influence of shared repertoire of the community when using ICTs.
7   There should be user-friendly language and graceful ways of bringing people into conversations.
8   Longevity – time is needed for communication and to build trust, rapport and a true sense of community.
9   Leadership is important to sustain the community. In the case of virtual CoPs, a moderator, facilitator or a list owner is important.

Knowledge creation and dissemination in VCoPs

Knowledge creation, sharing and learning take place through different modes in CoPs. Nonaka and Takeuchi (1995) introduced a four-stage knowledge transfer spiral model that explains the stages and forms of knowledge creation and transfer in an organisational context. These modes include socialisation, externalisation, combination and internalisation. Socialisation is where individuals acquire knowledge from others through shared experience, observation and imitation. Externalisation involves meaningful dialogues and reflections to articulate tacit knowledge into explicit concepts. It also includes a systematic collection and archiving of explicit concepts drawn from different sources for future use by the organisation. The utilisation of the archived knowledge to benefit when faced with a knowledge crisis is known as combination. Internalisation refers to the process of learning by doing and verbalising and documenting experiences. Hafeez and Alghata (2007) contextualised these modes of knowledge creation and sharing to virtual CoPs as summarised below.

- Socialisation – where knowledge creation and transfer take place through interactions with experts in virtual chat rooms and in seminars and workshops organised by the community. Also storytelling by CoP members to relate their experiences is a powerful communication tool which helps listeners form ideas and concepts.
- Externalisation – where a CoP holds a structured archive or repository that contains all the discussions that have taken place since the start of the CoP in a topic-by-topic structure.
- Combination – in the combination process, the structured archive that the CoP holds makes it possible for members to access information over a period of time and benefit from the use of 'CoP memory' if one is faced with a knowledge crisis situation.
- Internalisation – issuing electronic newsletters to promote events, courses, publications, stories and ideas help in the internalisation of knowledge within the community.

Developing a virtual community of safety practice (VCoSP) for construction

Terms of reference

Flannery and Hinze (2008) suggested some valuable terms of reference for a virtual community of safety practice in construction as listed below.

1 In the spirit of sharing safety information freely among all interested parties, the membership to the community should be open to any individuals who are interested in construction OHS. This may include professionals from different types of contractors, designers, researchers and OHS authorities.
2 A query process is central to the working of the community of safety practice, whereby members can post questions for comments with inputs being provided by other members.

3   It could benefit the construction industry to a greater extent if the work products of the community of safety practice can be viewed by the public who can pose questions, because these could lead to significant changes in the overall performance of the construction industry. Adequate efforts should therefore be made to publicise the existence of the community.

From a functionalities point of view, Schweitzer (2003), Molphy *et al.* (2007) and Gannon-Leary and Fontainha (2007) listed important features that should be included in the virtual community of safety practice platform.

1   Personal identity (virtual business card) tool for creating personal identity or individual biography pages as a way to give contextual information about people. Community members are much more likely to engage in social interactions if they know about the individual they are engaging.
2   Discussion boards and email groups for asynchronous threaded discussions.
3   Chat and conferencing tools for synchronous discussions.
4   Tools for circulating community news among its members.
5   Community resources archives/repositories to retain the work products of the community over time.
6   Document exchange tools for sharing documents, photos, video clips and mp3 sound.
7   Community event manager tools for facilitating face-to-face meetings and forums for community members.
8   Quiz/survey/poll tools for getting the views of community members on various issues.
9   Blog/notes tools for letting individuals share OHS experiences with other members in the VCoSP through written entries.
10 CoPs should provide a way of searching through their site materials such as documents, blogs, stored threaded discussions, etc.
11 Section-based and taxonomy-based navigations are important.

Conceptual model of the VCoSP

A conceptual model for the virtual community of construction safety practice (VCoSP) was developed, as shown in Figure 5.2, to address the requirements mentioned in the terms of reference above.

There are four key components in the proposed model for the VCoSP:

1   Online interactions between members of VCoSP – these take place in different forms, such as seeking information through enquiries, initiating threaded discussions on topics of concern, sharing success stories of applying innovative ideas and conducting online forums and brainstorming sessions.

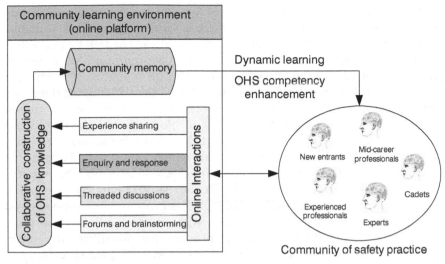

Figure 5.2 Conceptual model of VCoSP

2 Collaborative construction of new knowledge – by collaborating through various modes of interaction as mentioned above, new OHS knowledge is created collaboratively by the members of VCoSP and validated collaboratively by them before it is applied on site. The consultation and threaded discussion sessions function to perform the validation exercise of new knowledge automatically.

3 Retaining new OHS knowledge for reuse – new knowledge created by collaborations of members is a valuable asset that adds value to organisational practices and thereby improves competitiveness. Hence, this asset should be retained for reuse in the future. The community memory functions as the repository to store this asset in an organised manner.

4 Dynamic learning and competency enhancement – learning among the community members occurs in two modes. The first mode is where members who participate in the interaction sessions as well as observers learn as new knowledge is created. In the second mode, members can retrieve retained knowledge from the community memory when they are struck by a need to learn. Both of these modes facilitate OHS competency advancement for individual members. And learning via these modes takes place continually as a dynamic process.

Developing the online system

There were two possible options to develop the VCoSP, including: (1) coding using web development languages such as HTML, PHP, Cold Fusion, ASP, JSP, etc., or (2) using a Contents Management System (CMS). It was decided to use a CMS for this task on account of its many benefits that have already been explored in previous chapters. There is a wide range of CMS available, both commercially and free general public licences. The free CMS are also called Open Source CMS. The selection of an appropriate CMS is heavily dependent upon the functionalities available in a particular CMS and those required from a developed system. Given the social networking and knowledge sharing nature of the proposed VCoSP, WordPress was chosen for the development task. WordPress is a blogging tool and publishing platform powered by PHP and MySQL. WordPress, together with BuddyPress, allows the creation of a social networking site similar in use to Facebook. WordPress and BuddyPress are Open Source, available for free from their respective websites. While WordPress was originally designed as a system for blogging (started in 2003), it has evolved into a sophisticated content management system and more, with an enormous range of plug-ins available. BuddyPress is a younger product than WordPress, developed in 2008, with a first stable release in 2009. It was designed specifically to add social networking features to a WordPress powered site. Being a web development tool, WordPress, together with BuddyPress, had to be installed on a web server and the development task also had to be performed on the server. Server space was leased from a commercial provider and the development was done directly on it.

Creating the VCoSP in BuddyPress

The creation of the virtual community of safety practice involved the following steps.

1   Downloading WordPress from the wordpress.org site, and BuddyPress from buddypress.org.
2   Installing WordPress on the web server. The process is relatively easy and there is a large set of information guides on the WordPress site to help with the process.
3   Installing BuddyPress as a plug-in to WordPress.
4   Modifying the BuddyPress theme for a desired interface and look and feel.
5   Installation of a number of BuddyPress plug-ins to extend the functionality. These include:
    • BP Activity Plus, a free plug-in from WordPress MU, allows users to post images, video, and links in the Activity/Wall stream.
    • BP Group Documents, a free plug-in from WordPress MU, allows members to upload and share files between members of a group.
    • BP Album, a free plug-in from Wordpress.org, that mimics Facebook Photo application.
    • BP Chat, a free plug-in from BuddyDev, provides a chat feature similar to that in Facebook.

The end result was a site that contains:

- Members space – a members' profile, activity stream (like Facebook's wall), messages (allowing internal emailing amongst members), friends list and an album space.
- Groups space – members can create their own group or join existing ones. Groups have their own activity stream, forum and document sharing.

Both the Member and Group spaces have administrative controls, covering settings from personal information and passwords to visibility of group forums to members. Figure 5.3a represents the main (log-in) window and interaction windows of the online platform.

Member collaboration and safety knowledge management

The virtual platform features five key tools for member collaboration and safety knowledge management and they appear as sidebar tabs, as shown in Figure 5.3b and c.

1   *Messages*: the Messages tab is the central location for recent information posted by community members and it is where up-to-date contents from community activities are kept. When a community member shares content (videos, photos, notes, etc.), they will appear under Messages for high visibility to other members.
2   *Forums*: the Forums board enables community members to initiate and engage in threaded discussions about topics that are of interest at a given time, share experiences with other members in the community through written entries (other members can also write comments on entries) and ask other members for information. Likewise, members can share anything on the internet by posting the link onto the community platform. Members can post websites, blogs, YouTube videos, pieces of news, etc.
3   *Album*: With the Album tab, members can upload pictures to the online platform and define visibility restrictions and other descriptions/notes. Other members can comment on those photos as appropriately.
4   *Events*: The Events tab helps promote community events. It allows the creating of an event, adding pictures to suit the theme of the event and inviting community members to participate. Also, members can RSVP. There are also options to leave the event open to members in the community alone, invite guests and/or open to everyone.
5   *Chat*: Chat is an instant messaging tool. The chat tool is available at the bottom of the VCoSP page and it lists members who are online. With Chat, community members can have synchronous discussions.

Apart from these user-centred functionalities, the virtual platform also features back-end administrative functionalities to constantly manage the community's privacy, contents, membership, access level and mass messaging.

Figure 5.3 VCoSP interfaces

Industry evaluation of VCoSP

An industry evaluation of the proposed VCoSP was undertaken in order to assess the implications of the proposed system for builders and challenges facing its implementation. A case study approach, supplemented by a focus group, was utilised for this purpose. The details of the evaluation exercise and its findings are described below.

Case study and focus group

The proposed system was implemented in a small-sized specialised contractor organisation. For confidentiality reasons, the name of the organisation is not disclosed but anonymously identified as RBG. The organisation carries out some of the high hazardous trades in construction projects. The types of trades that RGB undertakes include:

- bulk earthworks and foundation construction;
- demolition works; and
- contaminated land management.

RBG possess a decent pool of permanent resources, including both professionals and plant. The pool of professionals includes project managers, engineers, estimators, surveyors, supervisors, leading hands, plant operators and labourers.

Although, RBG has been undertaking high-risk jobs, they did not have online systems to promote safety on their sites. When the author approached them for evaluation, they understood the potential benefits that the proposed system could offer and accepted the invitation without a second thought. The VCoSP was implemented for RBG and its employees were encouraged by the managing director to become members of the online community. Due to confidentiality reasons, screenshots of RBG-specific VCoSP are not included in this chapter. After RBG had used the VCoSP for a period of fifteen months, a focus group was administered by the author with the employees of RBG with the aim of:

- assessing how the VCoSP has benefited RBG to improve its OHS performance; and
- identifying the challenges faced for the utilisation of the VCoSP.

Positive implications of the VCOSP

Focus group participants confirmed from their experiences of using the VCoSP that the following benefits could be enjoyed by a construction organisation if the VCoSP is set up and all site management team members within the organisation actively participate in its course:

1 Learning from success stories of 'how to' and stories of what has not worked.
2 Sharing innovative ideas and technologies that could assist and educate the field safety personnel to work more effectively in the management of safety.
3 Finding ways to positively influence safe behaviours.
4 Obtaining collaborative opinions from fellow experts on specific issues/ questions.
5 Collaborating and sharing strategies to address chronic challenges of diverse workforce, language barriers, skill shortages and poor safety attitude of workers.
6 Helping site professionals continually improve safety competency.
7 Identifying and standardising best safety practices for the organisation.
8 Improving productivity on site.
9 Staying informed of changes on regulations, policies and procedures concerning safety that may affect obligations of safety professionals.
10 Meaningful changes to current practices might easily be implemented through the community of safety practice.
11 Aiding professionals to demonstrate safety leadership in the organisation.
12 Nurturing a strong safety culture within the organisation.
13 It is easy to warn and alert safety personnel on different sites of unexpected safety and hazardous concerns, such as health issues, radiation, epidemics, etc. on sites.
14 Knowing about leading indicators of safety flaws on site.
15 Getting directions to external sources of information, agencies and guidelines from community members.

Challenges for the successful implementation of the VCoSP

Despite many positive implications, focus group participants also indicated a few significant challenges that may impede the successful implementation of the VCoSP in an organisation. These include the following.

1 Weak motivation of individual project team members and other employees of an organisation to join the VCoSP and actively participate in its virtual interactions may lead the community to a dying state.
2 The VCoSP needs to work hard to maintain energy and a high degree of participation. Individual members of the community must engage with it in order that it may develop and grow.
3 The VCoSP lacks the opportunity for face-to-face interactions and socialising, which can consolidate group membership. Consequently, individuals may fail to engage in the VCoSP. Trust building is vital for sharing and trust primarily develops through face-to-face interactions. In the virtual environment, identities can remain hidden.
4 Poor computer skills of members.

Hence, it is paramount to put measures and strategies in place within an organisation to deal with these challenges so that the aforementioned benefits can be reaped.

Conclusions

Establishing a virtual community of safety practice within a construction organisation could offer a great deal of opportunity and potential for improving the occupational health and safety performance of the organisation. On one hand this could function as a platform for competency improvement and capacity building for the employees within the organisation, and on the other hand this could help uplift the safety standard on site. The use of social networking-enabled open source content management systems like BuddyPress provides a cost-effective technology for builders to set up a virtual community of safety practice with minimal administration efforts. The membership of VCoSP can also be extended to outside the organisational boundaries to establish an industry-wide community. This would create a strong safety mentorship culture within the industry and help address the chronic challenges facing the industry, including skill shortages and other competence-related causes of poor safety records.

6 Concluding remarks

Introduction

This chapter brings together all of the subthemes and findings in previous chapters into a single coherent model and reports its integrated applications in the construction industry. The chapter also discusses the practical implications of the findings, followed by key challenges for practical implementation in the industry. Finally, it sets further directions for research and explorations from the unique research arena that the book has instituted.

Summary

The construction industry is one of most dangerous industries globally. For example, the Australian construction industry records an incident rate of 22 per 1,000 employees, which is almost double the national rate for all industries in the economy. Compensation paid to injured workers amounts to 0.5 per cent of the total turnover of the Australian construction industry, which in turn increases construction costs by 8.5 per cent for clients. At the national level, construction accidents cost around 1.24 per cent of the GDP of Australia. Therefore, OHS has been a grave concern for policy-makers, authorities, builders, construction unions and researchers. In a bid to shake off this unpleasant aspect of the construction industry, numerous efforts have been made by these different bodies by way of: (1) establishing OHS regulations, codes of practice, standards and accreditation schemes, (2) providing different types of OHS training, (3) producing innovative OHS frameworks and tools and (4) strictly implementing safety management systems on site. In spite of these concerted efforts, the incident rate in construction still remains at an unacceptably high level. Studies suggest that the construction industry is characterised by numerous unique features that make OHS management and accident prevention quite challenging for builders. Among the unique features are the following.

- The construction process is dynamic and complex as activities, workforce mixes, materials and equipment used, site layout and activity interfaces change constantly over the course of construction. It is quite challenging for the site team to have strict control over dynamic hazardous conditions associated with many concurrent non-routine jobs on a constantly changing site set-up.

- Pressure for greater efficiency, exerted by tight budgetary and schedule constraints, results in disregarding safety and encouraging hazardous practices and unsafe behaviours. It also discourages organisations from investing in safety.
- There is a serious mismatch of OHS competencies amongst construction professionals due to high employee turnover and globalised job market systems. This causes a significant challenge to effective OHS management and thus demands for a continual safety training system to be in place for less competent professionals. On the other hand, it is difficult for them to attend workshops and training owing to tight project schedules and work pressure.
- A multicultural migrant worker demographic is one of the defining characteristics of the construction industry today. Migrant workers who come from countries where OHS standards are relaxed have a different perception of risk, and bring with them dangerous work practices from their countries of origin. Also language barriers amongst migrant workers have direct impacts on their safety. Their poor training and inability to understand basic safety instructions, warning signs and engage in safety communications expose themselves and others to greater risks.
- A chain of subcontractors is commonly observed in construction owing to the diversity of activities. Construction sites with multiple subcontractors make monitoring and enforcement of OHS more difficult for head contractors, and increases the chance of OHS non-compliance escaping undetected.
- The poor safety performance of the construction industry is attributable to a lack of commitment to safety from senior management to a large extent. However, owing to the combined effect of the fragmented nature of construction projects and work pressure on top management, it seems difficult for them to personally be involved in and monitor site safety affairs on a regular basis.
- Continual improvement in OHS, led by innovation and learning from past experiences, is heavily hampered by the very nature of the construction industry. The fragmented nature of construction, coupled with each project having its own schedule and work pressure constraints, eliminates opportunities for social learning of OHS by employees.

Given the unique and diverse nature of these challenges, this book synthesised a novel strategy to overcome them in improving safety in construction projects. A knowledge-based OHS management strategy was advocated to underpin the OHS management process with new capabilities of: (1) being site boundary independent, (2) actively involving site staff, subcontractors, workers and head office safety staff, (3) learning and improving from experiences of site teams and incidents in projects and (4) continually incorporating innovations in safety management. Via a tactful synergy of knowledge management concepts, OHS principles and web technologies, four web-based systems, namely a corporate OHS knowledgebase, e-safety trainer, virtual community of safety practice and project OHS diary, were developed and validated with potential end-users through sound research methods. Brief descriptions of these four systems and their functionalities are given below.

Corporate OHS knowledgebase

The ability to identify health and safety hazards as early as possible and implement good control measures is vital to a project of any size. Pre-construction OHS planning is identified as a critical step in that process. It is also a core requirement in most OHS regulations globally. Nonetheless, developing effective OHS plans faces significant challenges, including the large scopes of construction projects, skill shortages among construction professionals and abundance of OHS knowledge in unorganised formats. A web-based OHS knowledgebase was developed as a way of addressing these challenges and thereby assisting in effective OHS planning. The system was then evaluated by its potential end-users. The evaluation results suggest that the web-based OHS knowledgebase: (1) captures OHS knowledge from different sources and retains it in a single and easily accessible virtual location, (2) is capable of providing on-demand OHS planning information, (3) could help site staff with on-the-job learning of OHS skills and (4) could help minimise accidents on site and thereby save time and money for builders.

E-safety trainer

Unsafe behaviours of workers have been repetitive causes of accidents in construction. Current OHS training schemes for workers that are being used by the construction industry seem not to have achieved significant success in dealing with this issue. Unsafe behaviours can be changed by changing workers' attitudes towards hazards and safety. Affective education and training influences learners' attitudes, values and motivation. Findings in neurosciences, psychology and cognitive sciences also recognise that affect plays a crucial role in guiding rational behaviour, memory retrieval, motivation, attention, decision-making and creativity. The deployment of a mentally influential medium is crucial to achieve the desired attitudinal change in workers. In this vein, an online system was developed to provide interactive training resources, which facilitate affective OHS training to construction workers in site inductions and other OHS training programmes. The new e-tool was evaluated by builders and the findings suggest that the tool can have many positive implications for construction workers and builders, including: (1) constant utilisation of interactive learning resources in the e-tool in site inductions, tool box talks and other OHS training programmes could change even hardened negative attitudes in workers and (2) the e-tool can help builders to prove OHS training to workers on their sites, irrespective of their location and remoteness from the main office of the builders.

Project OHS diary

The effective implementation and monitoring of OHS management systems on construction sites is critical to prevention of accidents. However, there are many practical challenges facing this on site, including: constantly changing hazardous conditions as a project progresses, multilayered subcontracting and losing track of OHS responsibilities, concurrent progresses of multiple tasks on site, abundance of hazardous material use, plant operations and tradesmen, limitations of supervisors in monitoring every aspect meticulously, and a large amount of paperwork required and the risk of losing records. A groupware system was developed to help effective implementation and monitoring of OHS management systems in construction projects. An evaluation exercise of the groupware was administered with construction site professionals. The results indicate that the groupware system is able to deliver numerous benefits to builders, notably: (1) real-time capture of OHS implementation data from sites helps track actions and measure progress on critical health and safety actions, (2) with effective record-keeping practices, builders can easily demonstrate their OHS compliance and avoid penalties and prosecutions by OHS authorities, (3) organised OHS data capturing, processing and communicating enable OHS performance measurement and continual improvement and (4) organisational learning of OHS from mistakes and incidents is fostered, leading to innovation in OHS practices.

Virtual community of safety practice

Communities of practice provide a platform for sharing know-how and experiences amongst employees from different organisations. They could offer a great deal of opportunity and potential for improving the occupational health and safety performance of builders. However, the application of the concept of communities of practice in the construction industry faces a major challenge. Because construction projects are remote and scattered, and each project has its set completion schedule and progress pressure, interactions between employees of different organisations is difficult. It is impractical to establish a group of OHS practitioners and other construction professionals from different construction sites or organisations who could meet on a regular basis to share safety-related experiences and knowledge. Nonetheless, the formation of online communities of practice, departing from the conventional model, would alleviate the impediments caused by geographical, time and work pressure constraints. A virtual community of safety practice (VCoSP) for construction organisations, leveraging on the power of Web 2.0 technologies, was developed and validated through a case study in a construction organisation. The validation exercise reinforced that the implementation of the VCoSP within a construction organisation could bring about numerous benefits, including: nurturing a strong safety culture within the organisation and helping site professionals continually improve safety competency.

Integrated OHS management portal

Although the four web-based systems above have been discussed in an independent fashion, they can be integrated into a single portal to form a comprehensive web-based infrastructure for knowledge management enabled OHS management within a construction organisation. Figure 6.1 shows a model arrangement of the various systems within a single portal. The four web systems may be housed in a single server space with an independent database and a subdomain for each system. This would provide flexibilities in the implementation and maintenance of the portal as defined below.

* Different web development technologies may be used for different subsystems. As described in the preceding chapters, each system in the book has been developed using a different web development method. When a separate database is used for each subsystem, it would enable the benefits of using multiple technologies to be reaped.
* Implementation and maintenance work may be carried out on a modular basis.
* In the event of a failure of one subsystem in the portal, other subsystems would still be usable.

Practical implications

The research findings reported in this book have significant implications for the construction industry, the ICT industry and the national economy, as expounded below.

Implications for the construction industry

The book pictures a new shape to OHS management in construction through its online approach, underpinned by knowledge management principles. The construction industry players declared that the new approach could deliver numerous benefits for their industry, notably:

* improving safety and reducing incidents in the construction industry,
* improving OHS compliance by builders and thereby avoidance of penalties and prosecutions that damage business reputation,

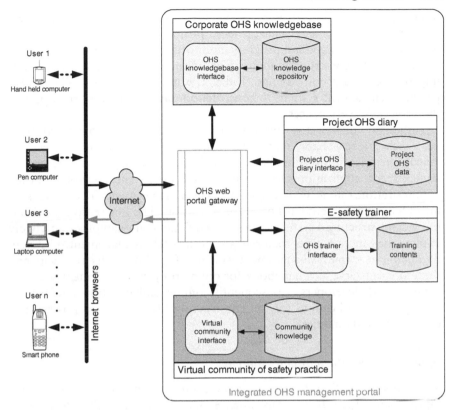

Figure 6.1 Integrated OHS management portal

- facilitating on-the-job continual learning of OHS for construction site team members,
- promoting a strong safety culture within the construction industry, and
- planting the seed for organisational learning in construction OHS, paving the way to innovation in OHS practices.

Implications for the ICT industry

Construction projects are often delivered by multi-disciplinary, multi-company and multi-location groups of people. Timely communications and exchange of information and documents between these different companies are crucial for project success. The construction industry was prone to delays, cost overruns and inconsistencies in information. The ICT industry provided a solution to improve the situation. Beginning from late 1990, online collaboration and project management applications, also known as project websites, project extranets or construction groupware, were introduced into the construction industry. These systems integrate multiple parties to a project through online collaboration platforms and facilitate document management, project communications, project workflow management and progress monitoring. With the adoption of this technology, the construction industry has witnessed significant productivity improvements and cost savings. Today, the usage of online collaborative project management systems in construction has become a normal part of the project delivery process.

Although collaborative project management using project extranets has reached a reasonably mature state in construction, its uptake in OHS management is still in its infancy. The main focus of existing online project management applications has been around document control, contract administration, online tendering and workflow management. Their extension into OHS management has not yet seen fruition. The key reason for this is the lack of structured system analysis documents and examples that vividly explain the functionalities required within the OHS management subsystem of an online project management system. This book fills this gap and therefore can serve as a handbook for online project management software vendors and developers. Consequently, online collaboration and project management applications for the construction industry could possess new dimensions and directions.

Implications for the national economy

Safe Work Australia estimated that the total economic cost of workplace injuries and illness to the Australian economy for the 2005/06 financial year was $57.5 billion, representing 5.9 per cent of GDP. The construction industry alone represented 21.6 per cent of this cost and about 1.24 per cent of GDP (Safe Work Australia 2010). Compensation costs for all injuries in the construction industry amount to 0.5 per cent of the Australian construction industry's turnover, which is increasing building production cost by up to 8.5 per cent (McKenzie and Loosemore 1997; ASCC 2009). Similar situations are evident in other countries too. Therefore, apart from reducing the needless human misery and suffering associated with workplace injuries and illness, the research findings reported in the book also contribute directly to reducing economic costs and lost productivity for builders, and the cost of construction for building clients. It has been recognised repeatedly that improved safety in construction will have a major positive impact on the prosperity and competitiveness of a country's economy as a whole. Since construction is one of the largest industries in an economy, the potential economic and social benefits of the research findings are enormous. For example, the Australian construction industry accounts for about 7.25 per cent of GDP (about $82 billion per year) and directly employs about 950,000 people (9 per cent of the Australian workforce) (Safe Work Australia 2010); contributing to the well-being of 9 per cent of the workforce is significant.

Challenges for implementation

During the industry validations of the web-based systems discussed in this book, industry players highlighted some possible challenges for the successful implementation of the systems in the construction industry. Construction organisations intending to adopt the web-based infrastructure for knowledge management enabled OHS management may encounter any of them. Prior knowledge of these potential challenges will help the organisations to devise sound implementation plans to overcome these possible challenges smoothly. Hence, the challenges are outlined below.

- Initial implementation of the online OHS management infrastructure and training to staff may attract some expenses. This may discourage some builders, especially at times of low demand for construction.
- There may be a percentage of construction managers and site supervisors who are not computer literate, apprehensive about new technologies or/ and do not realise the benefits the system could provide. Usually, they tend to trust traditional methods more than innovative approaches. This could eventually turn into resistance to changing the traditional methods and adopting the proposed systems.
- Internet connection quality in remote sites is usually quite poor, which may cause obstacles to online management of OHS.
- Updating contents of the corporate OHS knowledgebase and the OHS training courseware in pace with legislative changes is crucial but it may be costly for builders.
- Small subcontractors may not have access to computers and internet connections in the workplace. In such circumstances, full utilisation of the system may not be practical.

Further directions

Developing new strategies and solutions to improve the poor OHS performance of the construction industry has been a main focus for policy-makers, OHS authorities, builders and researchers. Various solutions and strategies are being advocated by the different interested bodies. This book advocated the use of knowledge management concepts to revamp the OHS management process. The focus of the discussions has been the interest of builders. Nonetheless, there are many other potential applications of knowledge management concepts in OHS management that may benefit designers too.

Research confirms that design is a significant causal factor in fatalities and serious injuries in the construction industry. For example, a detailed review of 100 construction accidents in the UK found that, in 47 per cent of cases, a design change would have reduced the risk of injury (Gibb *et al.* 2004). Similarly, an analysis of 450 OHS incident reports in the US found that, in 30 per cent of the cases, the hazards that contributed to the incidents could have been eliminated if prevention through design measures had been implemented (Behm 2005). In Australia too, Creaser (2008) confirmed that 37 per cent of workplace fatalities in construction involved design-related causes. Given the important role that design decisions have in creating OHS risks, one of the most effective means of mitigation is to eliminate them at source. Many governments around the world have therefore mandated accident 'Prevention through Design' (PtD) in their OHS legislation. Designers are now required by law to thoroughly assess OHS risks in each design component as they make design decisions and seek to eliminate or reduce the risks. Nonetheless, while PtD is welcomed as an effective way to design out OHS risks early in a project, it faces a crucial implementation challenge. Industry surveys show that designers do not always have the knowledge and skills necessary to undertake PtD effectively, exposing workers to unnecessary risks and designers to new legal liabilities (Cooke *et al.* 2008; Zou *et al.* 2009). Design firms too face challenges in providing professional development training for their employees due to lean profit margins and heavy workload that prevent them from releasing their employees (Dingsdag *et al.* 2006; Ho and Dzeng 2010). Knowledge management can play a vital role in addressing deficiencies in PtD skills and competencies of designers. To this end, future research studies may be carried out to develop PtD knowledge management systems to enable designers to acquire PtD competencies with increased flexibilities in time, cost and location, and thereby design less risky to construct buildings.

References

Abdelhamid, T. S., and Everett, J. G. (2000) Identifying root causes of construction accidents. *Journal of Construction Engineering and Management*, 126(1): 52–60.

Abudayyeh, O., Fredericks, T. K., Butt, S. E., and Shaar, A. (2006) An investigation of management's commitment to construction safety. *International Journal of Project Management*, 24(2): 167–74.

ACIL (1996) *The Residential Subcontract Sector Canberra*, Canberra: Australian Government Publishing Service.

Ahmad, R. K., and Gibb, A. G. F. (2003) Measuring safety culture with SPMT – Field Data. *Journal of Construction Research*, 4(1): 29–44.

Al-Ghassani, A., Anumba, C. J., Carrillo, P. M. and Robinson, H. S. (2005) Tools and techniques for knowledge management. In Anumba, C. J., Egbu, C., and Carrillo, P. (eds), *Knowledge Management in Construction*, Oxford: Blackwell Publishing, 83–102.

Anumba, C. J., and Ruikar, K. (2008) *E-Business in Construction*. Brighton: Wiley-Blackwell.

Australian Safety and Compensation Council (ASCC) (2007) *National code of practice for induction for construction work*. Canberra: Commonwealth of Australia.

Australian Safety and Compensation Council (2009) Compendium of workers' compensation statistics Australia 2006–07, URL (accessed Jan. 2012): http://www.safeworkaustralia.gov.au/NR/rdonlyres/6661279D-D04E-4142-97E7-69997CEC0157/0/PARTEJourneyclaims200607.pdf.

Behm, M. (2005) Linking construction fatalities to the design for construction safety concept. *Safety Science*, 43(8): 589–611.

Borden, C. W. (1986) Hazards of double-deck single post shoring systems, URL (accessed Aug. 2009): www.imgs.ebuild.com/woc/c860713.pdf.

Bureau of Labor Statistics (2010) Census of fatal occupational injuries summary 2009, Economic News Release, URL (accessed Feb. 2012): www.bls.gov/iif/oshcfoi1.htm.

Campbell, J. M., Smith, S. D., Forde, M. C., and Ladd, R. D. (2007) Identifying hazards in transportation construction and maintenance tasks: a case-based reasoning approach using railroad data, URL (accessed Feb. 2010): http://www.see.ed.ac.uk/IIE/research/ndtman/pubs/TRB07-0147.pdf.

Carlsson, S. A. (2003) Knowledge managing and knowledge management in inter-organisational networks. *Knowledge and Process Management*, 10(3): 194–206.

Carter, G., and Smith, S. D. (2006) Safety hazard identification on construction projects. *Journal of Construction Engineering and Management*, 132(2): 197–205.

Charles, M., Pillay, J., and Ryan, R. (2007) *Guide to Best Practice for Safer Construction: Literature Review 'from Concept to Completion'*, URL (accessed January 2011): http://eprints.qut.edu.au/27637/10/7j._Guide_to_Best_Practice_for_Safer_Construction_Literature_Review.pdf.

Cheung, S. O., Cheung, K. K. W., and Suen, H. C. H. (2004) CSHM: Web-based safety and health monitoring system for construction management. *Journal of Safety Research*, 35(2): 159–70.

Chua, D. K. H., and Goh, Y.M. (2004) Incident causation model for improving feedback of safety knowledge. *Journal of Construction Engineering and Management*, 130(4): 542–51.

Ciurana, E. (2009) *Developing with Google App Engine*. URL (accessed March 2012): http://ebookee.org/Developing-with-Google-App-Engine-Firstpress_1949209.html.

Construction Safety Alliance (CSA) (2008) *OHSE SubbyPack*. Perth, Australia: Construction Safety Alliance.

Construction Safety Association of Ontario (CSAO) (2009) *Formwork*, URL (accessed Aug. 2008): http://www.csao.org/UploadFiles/Safety_Manual/Tools_and_Techniques/Formwork.pdf.

Construction Training Australia (2001) *Building and Construction Workforce 2006 (Draft 1): Strategic Initiatives*.

Cooke, T., Lingard, H., and Blismas, N. (2008) ToolSHeD™: the development and evaluation of a decision support tool for health and safety in construction design. *Engineering, Construction and Architectural Management*, 15(4): 336–51.

Creaser, W. (2008) Prevention through design (PtD) safe design from an Australian perspective. *Journal of Safety Research*, 39(2): 131–4.

Davenport, T., and Prusak, L. (1998) *Working Knowledge: How Organisations Manage What they Know,* Cambridge, MA: Harvard Business School Press.

Davis, N., and Gibb, A. (2009) Factors that increase health and safety risks for migrant construction workers. In Lingard, H., Cooke, T., and Turner, M. (eds), *Proceedings of the CIB W099 International Conference*, 21–23 October, Melbourne, CD-ROM, CIB W099, Paper No 32.

Debowski, S. (2006) *Knowledge Management,* Milton: John Wiley & Sons Australia.

Dekker, R., and Hoog, R. (2000) The monetary value of knowledge assets: a micro approach. *Expert Systems with Applications*, 18(2): 111–24.

Department of Commerce, the Government of Western Australia (2007) Formworkers and manual handling, URL (accessed Sept. 2009): http://www.docep.wa.gov.au/Worksafe/Content/Safety_Topics/ Manual_handling/Further_information/Formworkers_and_manual_handlin.html.

Department of Commerce, the Government of Western Australia (2010) *Hazardous Substances*, URL (accessed Oct. 2010): http://www.docep.wa.gov.au/WorkSafe/Content/Safety_Topics/Hazardous_substances/index.htm.

Department of Consumer and Employee Protection (2010) *Machinery and Equipment Safety: An Introduction*, URL (accessed Oct. 2010): http://www.docep.wa.gov.au/Worksafe/PDF/National_Standards/VWA_Machine_Safety_l.pdf.

Department of Immigration and Citizenship (DIAC) (2009) *Population Flows: Immigration Aspects – 2007,* Canberra: DIAC.

Dingsdag, D., Biggs, H., and Sheahan, V. (2006) *Safety Culture in the Construction Industry: Changing Behaviour through Enforcement and Education*, URL (acessed Oct. 2007): http://2006conference.crcci.info/docs/ CDProceedings/Proceedings/P132_Dingsdag_R.pdf.

Egbu, C. O. (2008) Knowledge management for improved construction e-business performance. In Anumba, C. J., and Ruikar, K. (eds), *E-Business in Construction*, Brighton: Wiley-Blackwell, 222–33.

Ekman, P., and Friesen, W.V. (1978). *Facial Action Coding System*, Palo Alto, CA: Consulting Psychologist Press.

Fahey, L., and Prusak, L. (1998) The eleven deadliest sins of knowledge management. *California Management Review*, 40(3): 265–76.

Fernie, S., Green, S. D., Weller, S. J., and Newcombe, R. (2003). Knowledge sharing: context, confusion and controversy. *International Journal of Project Management*, 21(3): 177–87.

Flannery, J., and Hinze, J. (2008) The value of a safety community of practice. *Proceedings of the CIB W099 International Conference on Evolution of and Directions in Construction Safety and Health*, 9–11 March, Gainesville, Florida, CIB W099, pp. 23–30.

Fong, P. S. W. (2003) Knowledge creation in multi-disciplinary project teams: an empirical study of the process and their dynamic interrelationships. *International Journal of Project Management*, 21(7): 479–86.

Gannon-Leary, P., and Fontainha, E. (2007) *Communities of Practice and Virtual Learning Communities: Benefits, Barriers and Success Factors*, URL (accessed April 2010): http://www.elearningeuropa.info/files/media/media13563.pdf.

Gano-Phillips, S. (2010) *Affective Education and Experiential Learning*, URL (accessed Feb. 2011): http://www.cuhk.edu.hk/oge/gesymposium/presentation/Susan_Affective.pdf.

Gherardi, S., and Nicolini, D (2000) The organisational learning of safety in communities of practice. *Journal of Management Enquiry*, 9(7): 7–18.

Gibb, A., Haslam, R., Hide, S., and Gyi, D. (2004) The role of design in accident causality. In Hecker, S., Gambatese, J., and Weinstein, M. (eds), *Designing for Safety and Health in Construction: Proceedings, Research and Practice Symposium*, Eugene, OR: UO Press.

Goldfayl, D. (1995) Affective and cognitive domain learning with multimedia: two sides of the same coin, ASCILITE Conference, Melbourne.

Hadikusumo, B. H. W., and Rowlinson, S. (2004) Capturing safety knowledge using design-for-safety-process tool. *Journal of Construction Engineering and Management*, 130(2): 281–9.

Hafeez, K., and Alghata, F. (2007) Knowledge management in a virtual community of practice using discourse analysis. *Electronic Journal of Knowledge Management*, 5(1): 29–42: www.ejkm.com.

Hanna, A. S. (1998) *Concrete Formwork Systems*, London: Taylor & Francis.

Haslam, R. A., Hide, S. A., Gibb, A. G. F., Gyi, D. E., Pavitt, T., Atkinson, S., and Duff, A. R. (2005) Contributing factors in construction accidents. *Applied Ergonomics*, 36(5): 401–15.

Health and Safety Executive (HSE) (2010a) *Permit-to-Work Systems*, URL (accessed Sept. 2010): http://www.powys.gov.uk/uploads/media/indg98_en.pdf.

Health and Safety Executive (2010b) Statistics on fatal injuries in the workplace 2009/10, URL (accessed Jan. 2012): http://www.hse.gov.uk/statistics/fatalinjuries.htm.

Hennink M., Hutter, I., and Bailey, A. (2011) *Qualitative Research Methods*, London: SAGE.

Herreid, C. F. (1997) What makes a good case? *Journal of College Science Teaching*, 27(3): 163–5.

Hinton, C. M. (2002) Towards a pattern language for information-centered business change. *International Journal of Information Management*, 22(5): 325–41.

Hinze, J., and Gambatese, J. (2003) Factors that influence safety performance of specialty contractors. *Journal of Construction Engineering and Management*, 129(2): 159–64.

Hinze, J., Huang, X., and Terry, L. (2005) The nature of struck-by accidents. *Journal of Construction Engineering and Management*, 131(2): 262–8.

Hinze, J., Pedersen, C., and Fredley, J. (1998) Identifying root causes of construction injuries, *Journal of Construction Engineering and Management*, 124(1): 67–71.

Hislop, R. (1999) *Construction Site Safety: A Guide for Managing Contractors*, Florida: Lewis Publishers.

Ho, C.L and Dzeng, R.J. (2010) Construction safety training via e-Learning: Learning effectiveness and user satisfaction. *Computers and Education*, 55(2): 858–867.

Holt, A. S. J. (2005) *Principles of Construction Safety*, Oxford: Blackwell.

Hu, K., Rahmandad, H., Smith-Jackson, T., and Winchester, W. (2011) Factors influencing the risk of falls in the construction industry: a review of the evidence. *Construction Management and Economics*, 25(3): 397–416.

Hubert, C., Newhouse, B., and Vestal, W. (2001) *Building and Sustaining Communities of Practice*, Houston, TX: American Productivity Centre.

Industrial Accident Prevention Association (IAPA) (2008) *Manual Materials Handling*, URL (accessed Sept. 2009): http://www.iapa.ca/pdf/manmat.pdf.

International Labour Organisation (ILO) (2010) *Training Package in occupational safety and health for the construction industry*, URL (accessed August 2010) http://www.ilo.org/sector/Resources/training-materials/WCMS_161706/lang--en/index.htm

Jonassen, D. H., and Hernandez-Serrano, J. (2002) Case-based reasoning and instructional design: using stories to support problem solving. *Educational Technology: Research and Development*, 50(2): 65–77.

Jones, A., and Issroff, K. (2005) Learning technologies: affective and social issues in computer supported collaborative learning. *Computers and Education*, 44(4): 395–408.

Jurewicz, R. A. (1988) *Worker Safety in Formwork Operations*, URL (accessed Sept. 2009): http://www.penoguard.com.au/images/stories/holes-in-construction.pdf.

Kamardeen, I. (2009) *Controlling Accidents and Insurers' Risks in Construction: A Fuzzy Knowledge-Based Approach*, New York: Nova Science Publishers.

Kartam, N. A. (1997) Integrating safety and health performance into construction CPM. *Journal of Construction Engineering and Management*, 123(2): 121–6.

Kimble, C., and Barlow, A. (2000) *Effective Virtual Teams through Communities of Practice*, Glasgow: University of Stratchclyde, Strathclyde Business School (Management Science Research Paper, 00/9).

Kirkpatrick, D. L. (1959) *Evaluating Training Programs*, 2nd edn, San Francisco: Berrett Koehler.

Kirkpatrick, D. L. (1998) *Another Look at Evaluating Training Programs*, Virginia: ASTD.

Kitts, B., Edvinsson, L., and Beding, T. (2001) Intellectual capital: from intangible assets to fitness landscapes. *Expert Systems with Applications*, 20(1): 35–50.

Koehn, E., and Datta, N. K. (2003) Quality, environmental, and health and safety management systems for construction engineering. *Journal of Construction Engineering and Management*, 129(5): 562–9.

Kort, B., Reilly, R., and Picard, R. W. (2001) An affective model of interplay between emotions and learning: reengineering educational pedagogy – building a learning companion. In *Proceedings of the IEEE International Conference on Advanced Learning Technologies* (pp. 43–6), Los Alamitos, CA: IEEE Computer Society Press.

Kovalchick, A. M., Hrabe, E., Julian, M. F., and Kinzie, M. B. (1999) ID case studies via the World Wide Web. In Ertmer, P. A., and Quinn, J. (eds), *The ID Casebook: Case Studies in Instructional Design* (pp. 141–8), Upper Saddle River, NJ: Merrill.

Kraiger, K., Ford, J. K., and Salas, E. (1993) Application of cognitive, skill-based and affective theories of learning outcomes to new methods of training evaluation. *Journal of Applied Psychology*, 78(2): 311–28.

Labour Force Survey (2009) *Safety and Health for the Construction Industry*, URL (accessed Oct. 2010): http://www.statistics.gov.uk/CCI/Nscl.asp?ID=5316&Pos =2&ColRank=1&Rank=80.

Laudon, K. C., and Laudon, J. P. (2002) *Essential of Management Information Systems*, 5th edn, Englewood Cliffs, NJ: Prentice Hall.

Lehtinen, S., Tammaru, E., Korpen, P., and Runkla, E. (2005) Information in occupational health and safety – bringing about impact in practice in Estonia, URL (accessed Aug. 2009): http://www.ttl.fi/NR/rdonlyres/2B3A4644-6B4F-484A-AFB3-508DF4FA7289/0/OccupationalHealthServicesinEstonia.pdf.

Liao, S. (2003) Knowledge management technologies and applications: literature review from 1995 to 2002. *Expert Systems with Applications*, 25(2): 155–64.

Liebowitz, J. (2005) Linking social network analysis with the analytic hierarchy process for knowledge mapping in organizations. *Journal of Knowledge Management*, 9(1): 76–86.

LifeLines Online (2010) *Planning for Case-Based Learning*, URL (accessed October 2010) http://www.bioquest.org/lifelines/PlanningStages.html

Lin, P. J. (2001) Using research-based cases to enhance prospective teachers' understanding of teaching mathematics and their reflections. Presented at The Netherlands and Taiwan Conference on Common Sense in Mathematics, Taiwan, Taipei.

Lingard, H., and Rowlinson, S. (2005) *Occupational Health and Safety in Construction Project Management*, New York: Spon Press.

Loosemore, M., and Andonakis, N. (2006) Subcontractor barriers to effective OHS compliance in the Australian construction industry. In Fang, D. P., Choudry, R. M., and Hinze, J. W. (eds), *Global Unity for Safety and Health in Construction: Proceedings of the CIB W99 International Conference* (pp. 61–8), Tsinghua University, Beijing: CIB.

Loosemore, M., Phua, F., Dunn, K., and Ozguc, U. (2009) Operative experiences of cultural diversity on Australian construction sites: implications for safety and well-being. In Lingard, H., Cooke, T., and Turner, M. (eds), *Working Together: Planning, Designing and Building a Healthy and Safe Industry. Proceedings of the CIB W099 International Conference*, 21–23 Oct.. Melbourne: CIB.

Love, P., Fong, P. S. W., and Irani, Z. (2005) *Management of Knowledge in Project Environment*, Oxford: Elsevier.

McDermott, R. (1999) How information technology inspired, but cannot deliver knowledge management. *California Management Review*, 41(4): 103–17.

McFadden, F. R., Hoffer, J. A., and Prescott, M. B. (2000) *Modern Database Management*, 5th edn, New York: Prentice-Hall.

McKenzie, J. and Loosemore, M. (1997) The value in health and safety in construction. In *Proceedings of the 13th Annual ARCOM Conference 15–17 September 1997*, King's College, Cambridge. Association of Researchers in Construction Management, Vol. 1, pp. 353–362.

Maddouri, M., Elloumi, S., and Jaoua, A. (1998) An incremental learning system for imprecise and uncertain knowledge discovery. *Journal of Information Sciences*, 109(1–4): 149–64.

Maine Municipal Association (MMA) (2005) *Best Practices Guide for Powered Industrial Truck 'Forklift' Safety*, URL (accessed Sept. 2009): http://www.memun. org/RMS/LC/bestprac/forklift.pdf.

Mayhew, C. (2002) OHS challenges in Australian small businesses: old problems and emerging risks. *Safety Science*, 6(1): 26–37.

Mel Crook & Associates (2010) *OHS and R: Site Safety Plan*, Sydney: Mel Crook & Associates.

Meyers, A. W. and Cohen, R. (1990) Cognitive-behavioural approaches to child psychopathology: Present status and future directions. In M. Lewis and S. M. Miller (eds), *Handbook of Developmental Psychopathology*, New York: Plenum Press, pp. 475–485.

Mládková, L. (2010) *Tacit Knowledge Sharing through Storytelling*, URL (accessed Jan. 2011): http://www.vgtu.lt/leidiniai/leidykla/BUS_AND_MANA_2010/Information_ and_Communication/0884-0890_Mladkova.pdf.

Molphy, M., Pocknee, C., and Young, T. (2007) *Online Communities of Practice: Are they Principled and How do they Work?* URL (accessed March 2010): http://www. ascilite.org.au/conferences/singapore07/procs/molphy.pdf.

Moridis, C., and Economides, A. A. (2008) Towards computer-aided affective learning systems: a literature review. *Journal of Educational Computing Research*, 39(4): 313–37.

Muller, W., and Wiederhold, E. (2002) Applying decision tree methodology for rules extraction under cognitive constraints. *European Journal of Operational Research*, 136(2): 282–9.

National Center for Case Study Teaching in Science (2010) *Teaching Resources*, URL (accessed April 2010): http://sciencecases.lib.buffalo.edu/cs/teaching/

Ng, T. S., Cheng, K. P., and Skitmore, M. R. (2005) A framework for evaluating the safety performance of construction contractors. *Building and Environment*, 40(10): 1347–55.

Nickols, F. (2003) *Communities of Practice: An Overview*, URL (accessed April 2010): http://www.providersedge.com/docs/km_articles/CoPOverview.pdf.

Nonaka, I., and Takeuchi, H. (1995) *The Knowledge Creating Company*, New York: Oxford University Press.

Occupational Safety and Health Administration (2003) *Personal Protective Equipment*, URL (accessed Aug. 2010): http://www.osha.gov/Publications/osha3151.pdf.

Oller, W., and Giardetti, J. R. (1999) *Images that Work: Creating Successful Messages in Marketing and High Stakes Communication*, Westport, CT: Greenwood.

Picard, R. W., Papert, S., Bender, W., Blumberg, B., Breazeal, C., Cavallo, D., Machover, T., Resnick, M., Roy, D., and Strohecker, C. (2004) Affective learning – a manifesto. *BT Technology Journal*, 22(4): 253–69.

Plutchik, R. (1980) A general psychoevolutionary theory of emotion. In Plutchik, R., and Kellerman, H. (eds), *Emotion: Theory, Research, and Experience*, vol. 1, *Theories of Emotion* (pp. 3–33), New York: Academic Press.

Port of Brisbane (2010) *Permit to Work Form*, Brisbane: Port of Brisbane, Health and Safety.

Queensland Government (2006) *Formwork Code of Practice 2006*, URL (accessed Sept. 2009): http://www.deir.qld.gov.au/workplace/resources/pdfs/formwork_code2006.pdf.

Quintas, P. R. (2005) The nature and dimensions of knowledge management. In Anumba, C., Egbu, C., and Carrillo, P. (eds), *Knowledge Management in Construction* (pp. 10–28), Oxford: Blackwell.

Racicot, B. M., and Wogalter, M. S. (1995) Effects of a video warning sign and social modeling on behavioral compliance. *Accident Analysis and Prevention*, 27(1): 57–64.

Ringen, K., Seegal, J., and Englund, A. (1995) Safety and Health in the Construction Industry. *Annual Review of Public Health*. 16(1): 165–88.

Robertson, J. (2003) *So, What is Content Management System?* URL (accessed Feb. 2010): http://www.steptwo.com.au/files/kmc_what.pdf.

Robinson, P. (2002) Call for safety shake-up. *The Age* (7 May), 7.

SafetyNet (2010) *OHS Consultation, Communication and Reporting*, URL (accessed May 2010): http://contractormanagement.com.au/OHS_consultation.html.

Safe Work Australia (2010) *Key Work Health and Safety Statistics, Australia*, URL (accessed Jan. 2012): http://www.safeworkaustralia.gov.au/NR/rdonlyres/75DC1241-76F2-4737-90F8-B75D3A2A78A7/0/Key_work_health_safety_statistics_2010.pdf.

Safe Work Australia (2011) *Construction Fact Sheet*, URL (accessed Jan. 2012): http://safeworkaustralia.gov.au/AboutSafeWorkAustralia/WhatWeDo/Publications/Pages/FS2010ConstructionInformationSheet.aspx.

Saurin, T. A., Formoso, C. T., and Guimaraes, L. B. M. (2004) Safety and production: an integrated planning and control model. *Construction Management and Economics*, 22(2): 159–69.

Schweitzer, S. (2003) *Functionalities of Online Communities of Practice*, URL (accessed April 2010): java.cs.vt.edu/public/classes/communities/.../schweitzer_project_draft.pdf.

Sheehan, T., Poole, D., Lyttle, I., and Egbu, C. O. (2005) Strategies and business case for knowledge management. In Anumba, C. J., Egbu, C., and Carrillo, P. (eds), *Knowledge Management in Construction* (pp. 50–64), Oxford: Blackwell.

Silberberg, R. (1991) *The Subcontractor and Australia's Housing Industry: An Example of World Class Competitiveness*. Keynote address: The Law and Labour Market, Proceedings of the Conference of the H. R. Nicholls Society, URL (accessed September 2010): http://www.hrnicholls.com.au/archives/vol11/vol11-1.php.

Snelson, C., and Elison-Bowers, P. (2009). Using YouTube videos to engage the affective domain in e-learning. In Méndez-Vilas, A., Martín, A. S., Mesa González, J. A., and Mesa González, J. (eds), *Research, Reflections and Innovations in Integrating ICT in Education* (vol. 3, pp. 1481–5), Badajoz: FORMATEX, URL (accessed Jan. 2011): http://www.formatex.org/micte2009/book/1481-1485.pdf.

Sommers, P. H. (1981) *Checklist for Safe Construction of Formwork*, URL (accessed Aug. 2009): www.imgs.ebuild.com/woc/c810111.pdf.

Sommerville, J., and Craig, N. (2006) *Implementing IT in Construction*, London: Taylor & Francis.

Sorine, A. J., and Walls, R. T. (1996) Safety management information system for decision making. *Professional Safety*, 41(11): 26–8.

Spielholz, P., Wiker, S. F., and Silverstein, B. A. (1998) An ergonomic characterization of work in concrete form construction. *American Industrial Hygiene Association Journal*, 59(9): 629–35.

Stricker, A. G. (2009) *Why Affective Learning in a Situated Place Matters for the Millennial Generation,* URL (accessed Jan. 2011): http://www.au.af.mil/au/awc/awcgate/a46/affective-learning-situated-place.pdf.

Tam, C. M., Fung, I. W. H., and Chan, A. P. C. (2001) Study of attitude changes in people after the implementation of a new safety management system: the supervision plan. *Construction Management and Economics*, 19(4): 393–403.

Tan, W. (2007) *Practical Research Methods*, 3rd ed. Singapore: Prentice Hall.

Teo, A. L., and Ling Y. Y. (2006) Developing a model to measure the effectiveness of safety management systems of construction sites. *Building and Environment*, 41(11): 1584–92.

Teo, E. A. L., Ling, F. Y. Y., and Chong, A. F. W. (2005) Framework for project managers to manage construction safety. *International Journal of Project Management*, 23(4): 329–41.

Tomkins, S. S. (1981) The quest for primary motives: biography and autobiography of an idea. *Journal of Personality and Social Psychology*, 41(2): 306–29.

Toohey, J., Borthwick, K,. and Archer, R. (2005) *OH&S in Australia: a Management Guide*, Melbourne: Thomson.

Toole, T. M. (2002) Construction site safety roles. *Journal of Construction Engineering and Management*, 128(3): 203–10.

Tooman, T. (2010) Affective learning: activities to promote values comprehension. *Soulstice Training,* URL (accessed Feb. 2011): http://www.soulsticetraining.com/commentary/affective.html.

Trajkovski, S., and Loosemore, M. (2005) *Safety Implications of Low-English Proficiency among Migrant Construction Site Operatives,* URL (accessed May 2008): http://www.cfmeu-construction-nsw.com.au/pdf/spsafetyimpllowenglprof.pdf.

Udeaja, C. E, Kamara, J. M., Carrillo, P. M., Anumba, C. J., Bouchlaghem, N. D., and Tan, H. C. (2008) A web-based prototype for live capture and reuse of construction project knowledge. *Automation in Construction*, 17(7): 839–51.

University of Western Australia (UWA) (2009) *Contractor Permit to Work,* URL (accessed Nov. 2010): http://www.safety.uwa.edu.au/__data/page/9730/Contractor_Permit_to_Work_Version_1_9_03082009_%282%29.pdf.

Wadick, P. (2005) Learning safety in the construction industry: a subcontractors' perspective. Unpublished M. Ed. (Hons) thesis, University of New England, Armidale, Australia.

Wadick, P. (2006) Learning safety: what next? The case for a learning circle approach. *Proceedings of the 9th Annual Conference, 19–21 April*, University of Wollongong, Australia, URL (accessed May 2006): www.avetra.org.au/annual_conference/documents/PreliminaryProgramTOTAL_000.pdf

Wadick, P. (2009) Adult education theory and learning safety: what next? The case for learning circle approach to training for workers with low literacy. In Lingard, H., Cooke, T., and Turner, M. (eds), *Proceedings of the CIB W099 International Conference*, 21–23 Oct., Melbourne, CD-ROM, CIB W099, Paper No 55.

Wang, G. (2005) Humanistic approach and affective factors in foreign language teaching. *Sino–US English Teaching*, 2(5): 1–5, URL (accessed Feb. 2011): http://www.linguist.org.cn/doc/su200505/su20050501.pdf.

Wenger, E. (1998) *Communities of Practice,* Cambridge: Cambridge University Press.

Wenger, E., McDermott, R., and Snyder, W. M. (2002) *Cultivating Communities of Practice: A Guide to Managing Knowledge,* Boston, MA: Harvard Business School Press.

Wild, B. (2005). Occupational health and safety – the caring client. In Brown, K., Hampson, K., and Brandon, P. (eds), *Clients Driving Construction Innovation: Mapping the Terrain* (pp. 22–39), Brisbane: Icon.Net.

Wilson, J. M. J., and Koehn, E. E. (2000) Safety management: problems encountered and recommended solutions. *Journal of Construction Engineering and Management,* 126(1): 77–9.

Wirtz, K. W. (2001) Strategies for transforming fine scale knowledge to management usability. *Marine Pollution Bulletin,* 43(7): 209–14.

Wong, M. L. (2001) A flexible knowledge discovery system using genetic programming and logic grammars. *Decision Support Systems,* 31(4): 405–28.

WorkCover NSW (1998) *Formwork Code of Practice 1998,* Sydney: WorkCover NSW.

WorkCover NSW (2001a) *Summary of the OHS Regulations 2001,* Sydney: WorkCover NSW.

WorkCover NSW (2001b) *Identification Tool for Formwork: Hazard Profile,* URL (accessed August 2009): http://www.workcover.nsw.gov.au/formspublications/publications/Documents/identific_tool_formwork_hazard_profile_0981.pdf.

WorkCover NSW (2001c). *Identification Tool for Steel Reinforcement Fixing: Hazard Profile,* URL (accessed August 2010): http://www.workcover.nsw.gov.au/formspublications/publications/Documents/identific_tool_steel_reinforcmt_fixing_hazard_profile_0983.pdf.

WorkCover NSW (2001d) *Identification Tool for Concrete Placement: Hazard Profile,* URL (accessed August 2010): http://www.workcover.nsw.gov.au/formspublications/publications/Documents/identific_tool_concrete_placement_hazard_profile_0984.pdf.

WorkCover NSW (2001e) *Identification Tool for Metal Roofing: Hazard Profile,* URL (accessed August 2010): http://www.workcover.nsw.gov.au/formspublications/publications/Documents/identific_tool_metal_roofing_hazard_profile_0978.pdf.

WorkSafeBC (2009a) *Safety Inspections,* URL (accessed Aug. 2010): http://www.worksafebc.com/publications/health_and_safety/by_topic/assets/pdf/safety_inspections.pdf.

WorkSafeBC (2009b) *Joint Occupational Health and Safety Committee,* URL (accessed Aug. 2010): http://www.worksafebc.com/publications/health_and_safety/by_topic/assets/pdf/jointoch.pdf.

WorkSafeBC (2010) *Toolbox Meeting Guide,* URL (accessed Aug. 2010): http://www.worksafebc.com/publications/health_and_safety/by_topic/assets/pdf/TG06-00_Checklist.pdf.

Zou, P., Yu, W., and Sun, A. C. S. (2009) An investigation of the viability of assessment of safety risks at design of building facilities in Australia. In *Proceedings of the CIB W099 Conference 2009,* 21–23 October, Melbourne, CD-ROM, CIB W099, Paper No 08.

Index

Page numbers in *italic* indicate Figures, Tables and Exhibits.

9 780367 380380